T0222420

NON-REPRESENTATIONAL THEORY

Non-representational Theory explores a range of ideas which have recently engaged geographers and have led to the development of an alternative approach to the conception, practice, and production of geographic knowledge. Non-representational Theory refers to a key body of work that has emerged in geography over the past two and a half decades that emphasizes the importance of practice, embodiment, materiality, and process to the ongoing formation of social life. This title offers the first sole-authored, accessible introduction to this work and its impact on geography.

Without being prescriptive the text provides a general explanation of what Non-representational Theory is. This includes discussion of the disciplinary context it emerged from, the key ideas and themes that characterise work associated with Non-representational Theory, and the theoretical points of reference that inspires it. The book then explores a series of conjunctions of 'Non-representational Theory and...', taking an area of geographic enquiry and exploring the impact Non-representational Theory has had on how it is researched and understood. This includes the relationships between Non-representational Theory and Practice, Affect, Materiality, Landscape, Performance, and Methods. Critiques of Non-representational Theory are also broached, including reflections on issues on identity, power, and difference.

The text draws together the work of a range of established and emerging scholars working on the development of non-representational theories, allowing scholars from geography and other disciplines to access and assess the animating potential of such work. This volume is essential reading for undergraduates and post-graduate students interested in the social, cultural, and political geographies of everyday living.

Paul Simpson is Associate Professor of Human Geography at the University of Plymouth where he is Co-leader of The Centre for Research in Environment and Society. His research focuses on the everyday life of urban spaces and often proceeds through a combination of ethnographic research and engagements with non-representational theories and post-phenomenological philosophies.

KEY IDEAS IN GEOGRAPHY

Series editors: *Noel Castree, University of Wollongong and Audrey Kobayashi, Queen's University*

The *Key Ideas in Geography* series will provide strong, original, and accessible texts on important spatial concepts for academics and students working in the fields of geography, sociology, and anthropology, as well as the interdisciplinary fields of urban and rural studies, development, and cultural studies. Each text will locate a key idea within its traditions of thought, provide grounds for understanding its various usages and meanings, and offer critical discussion of the contribution of relevant authors and thinkers.

MIGRATION, SECOND EDITION
Michael Samers and Michael Collyer

MOBILITY, SECOND EDITION
Peter Adey

CITY, SECOND EDITION
Phil Hubbard

RESILIENCE
Kevin Grove

POSTCOLONIALISM
Tariq Jazeel

NON-REPRESENTATIONAL THEORY
Paul Simpson

For more information about this series, please visit: www.routledge.com/series/KIG

NON-REPRESENTATIONAL THEORY

Paul Simpson

Routledge
Taylor & Francis Group

LONDON AND NEW YORK

First published 2021
by Routledge
2 Park Square, Milton Park, Abingdon, Oxon OX14 4RN

and by Routledge
605 Third Avenue, New York, NY 10017

Routledge is an imprint of the Taylor & Francis Group, an informa business

British Library Cataloguing-in-Publication Data
A catalogue record for this book is available from the British
Library

Library of Congress Cataloging-in-Publication Data
A catalog record has been requested for this book

ISBN 13: 978-1-138-55216-6 (hbk)
ISBN 13: 978-1-138-55219-7 (pbk)

Typeset in Joanna
by codeMantra

FOR AILSA

CONTENTS

FIGURES

BOXES

ACKNOWLEDGMENTS

This book has a fairly length history beyond its writing over the course of 2017–2019. It has been shaped by a wide range of encounters and events. I first encountered something called 'more-than-representational theory' via Hayden Lorimer early in 2005 in the School of Geographical and Earth Science at the University of Glasgow. That introduction, along with open-minded support of Chris Philo, set it all in motion. 'More-than' became 'Non-' later in 2005 when I moved to Bristol to take up an ESRC '1+3' studentship in the School of Geographical Sciences. There I completed the MSc Society and Space and a PhD supervised by JD Dewsbury. Those four years provided something of an apprenticeship in non-representational theories, particularly during all those Friday late-afternoons/early evenings spent in 'The Hawthornes' engaged in enthusiastic debate and discussion over a range of continental philosophy and social theory. I was very fortunate to be supervised by JD and to be amongst many other PhD students, and Faculty, who shared an interest in this. Of the former, Sam Kinsley, Jules Brigstocke, James Ash, Pepe Romanillos, Charlie Rolfe, and Jen Lea stand out. Of the latter, JD Dewsbury, Keith Bassett, Mark Jackson, Maria Fannin, and Emma Roe equally so.

I've split the time since Bristol between Keele University and the University of Plymouth. At Keele, I was lucky enough to work, and later collaborate, with Pete Adey. That encounter couldn't have been better timed in terms of helping me move off from the PhD. At Plymouth, I've been fortunate to be a part of a really collegiate department which has come to accept the sort of work I do despite it being variously described (good humoredly?) as: 'extreme' cultural geography; 'niche, even for cultural geography'; or simply 'wacky geographies'. Thanks to Jon Shaw, Richard Yarwood, Naomi Tyrrell, Federico Caprotti, Nicky Harmer, Jules Brigstocke (again), Mark Holton, Kim Ward, Matthew Rech, and a host of others for shaping that collegiate atmosphere. I've also been fortunate to be supported by a broader community of academics variously interested in non-representational theories and associated matters. That has included Ben Anderson, Paul Harrison, Mitch Rose, Derek McCormack, John Wylie, Pete Merriman, Amanda Rogers, Helen Wilson, Jonny Darling, David Bissell, Joe Gerlach, Thomas Jellis, Nina Williams, Damien Masson, Jean-Paul Thibaud, Sara Fregonese, to name just a few. Apologies to those others who I've inevitably forgotten here.

A significant portion of this book was drafted during a semester's sabbatical leave in Autumn-Winter 2017. I would like to thank the School of Geography, Earth, and Environmental Sciences at the University of Plymouth for granting and facilitating that leave. It'd be fair to say, though, that the book would have been finished sooner had I not been put at risk of compulsory redundancy during the following Autumn-Winter semester. There's a whole other book to be written about the trying affects and turbulent atmospheres that increasingly circulate in the higher education sector of the UK.

Thanks also to Tim Absalom for assistance with the production of several of the figures included here. Mark Jackson and John Hargreaves helped in gaining information about the MSc Society and Space at the University of Bristol and the School there more generally. And thanks to Derek McCormack for permission to use Figure 6.1. In addition, thanks go to Andrew Mould and Egle Zigaite at Routledge and Noel Castree for guiding me through the editorial process. The book benefited greatly from the comments of those who reviewed the initial proposal and, in particular, from the comments of the two anonymous manuscript reviewers.

Finally, thanks to Lou for putting up with me and Ailsa for learning to say 'I missed you, daddy' a week after I started writing this book.

INTRODUCTION

Prelude: a scene seen from an office window

It's 8.45am on a Monday morning in autumn. From the fifth floor window, I can see a crossing point over a main road in the center of a city that borders my University's campus. At the opposite side of the crossing is a large student accommodation village as well as a range of other residential and commercial premises. On this side, the University campus. The weather this morning has been inclement, a mix of sharp showers, gusty winds, and the occasional threat of the sun breaking through the clouds. The road is relatively busy with the tail end of rush hour. Pedestrians wait for the crossing signal to allow them to cross the road. Occasionally, someone runs across through a gap in traffic, provoking motorists to beep their horns. Others are sat on low walls or benches or are stood leaning against lampposts, checking phones while waiting on friends to arrive. A mix of University students and professionals approach the start of their day.

At first glance, there may not appear to be much of significance going on here. Looking again, but this time as a geographer shortly heading to give a lecture on 'What is Human Geography' to a group of recently arrived undergrads, there are a number of points that might start to garner attention and take on some significance. The small plastic English flags attached to the door frames of a passing taxicab might be noted and used to illustrate ideas of 'banal nationalism' which can, along with a host of other meaningful signs, come to pervade the cultural landscape (Billig 1995; Wylie 2007a). The transport modes being used here, their efficiency and/or sustainability (or not) (Shaw and Docherty 2014), or the meanings such movement holds (Adey 2017), might seem worthy of discussion. Alternatively, the status of this as a 'public space' might be a topic for discussion, asking questions about who is present, how the space is being used, and how the space is being overtly or tacitly managed. That might lead onto reflections on who is being excluded from use or access (Mitchell 2003).

Figure I.1 A scene seen from an office window (*Author's own*).

Furthermore, the estate agent's office to the right of the image might be flagged and questions asked about the balance of rental accommodation, student accommodation, and affordable housing for sale or rent. That could lead to discussions of gentrification and studentification given the impact of such developments on the surrounding area (Hubbard 2006; Smith and Hubbard 2014). These examples are by no means exhaustive, but would all be quite well-established topics for geographers to discuss in deconstructing such a scene that pick out aspects of its social, cultural, political, and economic geographies.

Thinking about this in another way, though, other points which might expand such a geographic take start to emerge. Drawing on a range of developments around what has been called 'non-representational theory', we might start to recognize that there are other things going on here that are less obviously seen but still potentially significant to the playing out of everyday life in this space.

For example, we might start to ask about things that are more felt in nature. I stood looking at this scene before heading to a staff meeting with a sense of enthusiasm and mild trepidation as another academic year started (that meeting, it turned out, rather dulled any such feeling of enthusiasm). In the scene itself, there are a whole host of bodies that are both encountering one another and are disposed in a range of ways. Some might have shared my feelings of enthusiasm, others feeling quite different. Some of them might be feeling happy and others sad. Some might feel anxious or excited, others hopeful or depressed, and others still sober, hungover, and/or medicated. This might be visible in their facial expressions, the tone of their conversations, their body language, or their gait. But it might also be less obviously visible but end up being something palpably present, even if loosely, in a sort of 'atmosphere' that seems to exist in the space. Some of those feelings will emerge from encounters happening there and then in this space as these people move through it together – people acknowledging each other, or deep in conversation, or getting in each other's ways. However, other aspects will have their origins and destinations in events taking place over longer durations. That might be the events of the previous evening, pending deadlines at University or work, concerns for people or events taking place some distance away, and so on.

Further, we might also start to ask questions about the role that this space and the broader environment it exists within plays here and specifically what choreographic functions it might possess when it comes to what these bodies do and how they feel. This space's layout, its design, its management, the specific weather conditions present therein, and so on can all play a part here. Traffic lights produce a certain stop–start rhythm interspersed by pedestrians crossing. This might sit ill at ease with other timetables – of work shifts, of lectures, of bus timetables, and so on – and so produce feelings of frustration and disruption within the routines of those moving through this space. A range of instructions are also dictated by various signage, some noted, some not, some actively shaping the mood of individuals, others barely registering. And, the occasional wind and rain means that hoods are put up, blocking peripheral vision and so creating a funneled gaze. This all plays a significant role in how this space appears to those who move through it at this time, the practices they engage in, and how that is experienced, whether it is realized by them at the time or not. The surroundings here may recede into the background as individuals excitedly talk to their friends about recent events. Equally, others may be painfully aware of every sound as their head thrums from the excess alcohol intake of the night before. Or tiredness may make them relatively indifferent to their surroundings, moving on auto-pilot and not taking anything in that is happening around them.

Furthermore, we might start to ask questions about the ways in which the situation of the people moving through this space relative to their surroundings is mediated by a whole range of technologies. For example, smart phones (potentially connected to headphones) shape those individuals' awareness of what is happening around them at any given time, as well as allowing them to be aware of things going on at some distance from their immediate surroundings. Such technologies, not to mention the marketing materials encountered here – adverts for club nights pasted to lampposts, fast food promotions on bus shelters and building sides, a stand operated by Jehovah's witnesses, amongst others – all exert a call for our attention. And this is very often a specific sort of call. A whole host of data is being collected here through, for example, smart phones on what catches the users' eyes, what links they follow, what they search for, and so on. All the time, algorithms process this data and increasingly tailor what images are projected to us.

In sum, these sorts of mundane, embodied, felt, interactive, unfolding, mediated encounters are precisely the stuff that geographers influenced by non-representational theories have been drawn to. There is a whole range of subtle, shifting, and elusive geographies at play here in the co-production of these individuals and the spaces they move through on a day-to-day basis. This book aims to provide an introduction to such geographies and to help the reader navigate their way through a host of geographers' work on them.

What is non-representational theory?

'Non-representational theory' ('NRT') has come to refer to a key set of ideas that has emerged over the past two decades or so in human geography. 'Non-representational thinking' was first introduced to geography by Nigel Thrift, a prominent geographer who spent a significant part of his academic career at the University of Bristol (see Chapter 1). While much of Thrift's early work focused on matters of political economy and temporality, a series of questions coalesced from this in terms of practice, agency, the relations of space and time, and the particular societal context in which such relations unfold (see Thrift 1996). Thrift initially turned to such non-representational thinking in an effort to develop an alternative approach to the conception, practice, and production of geographic knowledge. Put succinctly, such non-representational thinking sought to re-orientate geographic analyses beyond what was, at the time, perceived as an over-emphasis on representations (images, texts, and so on) and instead emphasize practice, embodiment, materiality, and process. Or in Thrift's (1996: 5) own words, he felt that:

A hardly problematized sphere of representation is allowed to take precedence over lived experience and materiality, usually as a series of images or texts which a theorist contemplatively deconstructs thus implicitly degrading practice.

By way of contrast, non-representational thinking was to draw attention to a host of "mundane practices, that shape the conduct of human beings towards others and themselves in particular sites" (Thrift 1997: 127). In this, the ultimate focus was to be on "the geography of what happens" (Thrift 2008: 2). In a little more detail, Lorimer (2005: 84) later suggested that what came to be collectively known as 'NRT' focused:

> on how life takes shape and gains expression in shared experiences, everyday routines, fleeting encounters, embodied movements, pre-cognitive triggers, practical skills, affective intensities, enduring urges, unexceptional interactions and sensuous dispositions … which escape from the established academic habit of striving to uncover meanings and values that apparently await our discovery, interpretation, judgment and ultimate representation.

Thrift's initial calls for a greater attention to practice, lived experience, materiality, and the like have been taken up by a range of geographers, and this thing called 'NRT' has since evolved in multiple directions (see Anderson and Harrison 2010a).

If you are reading this book, it is likely that you have encountered 'NRT' in one of a number of contexts. Student readers, for example, may have heard 'NRT' (or 'more-than representational theory' – see below) mentioned in a module on social and cultural geography. If so, it is likely that a range of other terms accompanied this – affect, embodiment, materiality, atmosphere, performance, performativity, and so on. Equally, 'NRT' may have featured in a module on 'The History of Geography' or 'Geographic Thought' or 'Space and Social Theory'. In which case, 'NRT' will likely have been presented as part of a long line of other 'turns' or 'paradigms' which have shaped the practice of geographic research over a wide timeframe. It's likely here that 'NRT' will have been contrasted to other approaches, perhaps most obviously 'New Cultural Geography' or The Cultural Turn. It may also be that 'NRT' is something you've seen referenced in a range of research publications or textbooks. Postgraduate or postdoctoral readers may have heard it mentioned at a conference, workshop, or in departmental seminars. It may be something your supervisor has suggested you read up on for dissertation research. Whatever the case, there's a fairly good chance that 'NRT' seems to be a bit elusive in terms of what 'it' refers to or what this 'theory' actually is. For those who've looked further, you would likely also have encountered not just the sorts of terms mentioned above but also reference to the work of a range of 'non-geographers', primarily philosophers and social theorists. Here, you might have read quotations from these figures that do not immediately make sense, and, if they do, their arguments might exceed what you'd normally call 'geography'.

There is a common theme that generally runs through such scenarios. 'NRT' is often felt to be difficult to grasp given the way that it mixes conceptual vocabularies, complex social theories, and references to seemingly esoteric continental philosophy; involves potentially unusual styles of research and writing; and, as there is often either a surprising empirical focus or as there isn't a clear empirical object of study at all. In the face of this potential confusion, this book aims to introduce this set of ideas, terms, and developments in a way that both makes them more accessible than might otherwise be the case and, in doing so, provide an orientation in pursuing these ideas further in your own writing and research.

In developing that introduction, it is worth pausing for a moment and saying a few things about the name itself: 'non-representational theory'. Initially, Thrift did not actually call it this. As Harrison (forthcoming) traces in some detail, initially, 'non-representational thinking' was suggested by Thrift as a source of inspiration existing outside geography. However, over a period of around a decade, we can see a shift in references from Thrift and others from 'thinking' to 'theories', from plural (theories) to singular (theory), and ultimately to the proper, capitalized name 'Non-representational Theory'. Hence 'Non-representational Theory' became 'a thing' – an approach with a project or agendas, a target of critique, a subject for textbooks and reference work entries (including this one). This name, proper and singular, itself has presented something of an impediment to the reception and understanding of work associated with that name and the potential contribution it makes to the practice of geographic research. There are at least two reasons for this.

First, the word 'non-representational' has been misunderstood, given a number of apparent connotations that it suggests. For some, the prefix 'non' has meant (or at least implied) a movement away from concerns with representations and text (Nash 2000). In this, it starts to be seen to be about being 'after-' or 'post-representational' (Castree and MacMillan 2004) or 'anti-representational' (see Jacobs and Nash 2003; Smith 2003). This impression is understandable. Some of the early outlines provided by Thrift were somewhat over-exuberant in their treatment of established trends in human geography at that time, especially given the way in which these outlines positioned themselves very strongly in opposition to 'New Cultural Geography'. This theme of 'NRT's' relationship with other strands of geographic thought will be returned to throughout this book (in particular, see Chapter 1). But, for now, it is important to note that such tensions did lead to the proposal of more inclusive nomenclature in an attempt to (re)build bridges between, for example, those working in or on 'NRT' and 'New Cultural Geography'. 'More-than representational theory' is now a fairly well-established alternative, having been suggested by Lorimer (2005) as a more inclusive name that would soften some of this oppositional tone. These debates aside though, as will be demonstrated throughout this book, work associated with 'NRT' is in fact interested in representations. Representation and representations have not been left behind. In such work, representations are considered for what they do in the unfolding of practices (Dewsbury et al. 2002). The 'force of representations' is seen in the way that they enter into relations and have capacities to affect and effect (Anderson 2019). This means that they are taken to be 'performative' and so play a part in the ongoing shaping of social life through the unfolding of various actions and interactions. Therefore, the critical target of the 'non' is not representations in and of themselves. Instead, the critical target is a specific way of thinking about the world – a form of 'representationalism' – which reduces the world to, and fixes and frames it within, text or discourse alone (Lorimer 2005). Under such representationalism, such texts and the interpretation of their content become the key agenda of geographic work. For much of the work collected under the name 'NRT', by contrast, it is practices that become the starting point for such work. In many ways, the alternative title suggested by Thrift (1997) that never seemed to stick – 'The Theory of Practices' – is more affirmative in orientation and so could have mitigated against such debate and critical reception.

Second, 'NRT' does not constitute an actual theory in the way that other prominent geographic theories do. For example, Central Place Theory articulates a general explanation which would allow geographers to predict and explain the size, number, and distribution of towns. As part of this, certain laws or tendencies are established, and there are certain assumptions that underlie this. Here, we have a singular idea and objective, a clear object of analysis, a clear set of data to be analyzed, and clear results to be presented – distribution can either be assessed (i.e. whether it is conforming or not) or proposed (a model of how it should be in the interest of efficiency) (see Cresswell 2013a). Work associated with 'NRT' doesn't really offer any of this. The closest we get to a singular idea in 'NRT' relates to the importance of practice both as a singular idea and as an object of analysis. However, practice has been understood in a range of ways here, and there is no clear agreement amongst those pursuing research informed by NRTs (contrast, for example, Dewsbury 2000; Harrison 2009).

Really, 'NRT' – if there is such a singular 'thing' – presents us with a style of thinking which values practice and process rather than a theory (Thrift 2008). In some ways, this means that it is better to think of 'NRT' in the plural, as non-representational theories (NRTs) (Lorimer 2008). Again, at the outset, much of the reference here was in the plural and was about 'thinking' rather than a 'theory'. 'NRT' really acts as an umbrella term for a wide range of ideas, concepts, theories, and approaches largely originating beyond the confines of geography which have in common concerns for practice. Within work done under this banner, ideas from post-structuralists, vitalists, phenomenologists, pragmatists, feminists, and a collection of relational and constructivist social theorists mix in varying concentrations and combinations producing quite diverse and at times seemingly contradictory accounts of this happening of the world (Anderson and Harrison 2010a). Again, this plurality can make 'NRT' as a singular thing hard to pin down. While this book retains a title in the singular as 'NRT' in reference to this 'thing' that requires some explanation, this book itself will attempt to draw out both some of NRTs' diversity throughout – highlighting various debates, discussions, and disagreement – but also show some of the continuities that occur within that. In general, from here on in, I will use the plural NRTs to reflect the plurality within this work. This might lead to a number of awkward grammatical constructions; it'd be much simpler if NRTs were the singular thing they are at times made out to be. However, at certain points, I will use 'NRT' (note the quotation marks) where that supposed 'thing' is the target of reference in the work being discussed.

Key themes

To draw out some of the commonalities amid the diversity in NRTs, I'm now going to give a brief overview of some of the central themes that recur amongst work developing NRTs and so lend NRTs some sense of identity or consistency. These introductions are brief and will be developed in more detail in the coming chapters. The following is really intended to act as a primer for what follows. These interrelated themes are process, subjectification, embodiment, affect, and agency.

Process

NRTs try to attend to the 'onflow' of everyday life (Thrift 2008). The world and the events that take place within it are taken to be dynamic, unfolding, and so based on processes. NRTs thus openly acknowledge the partial and incomplete nature of the accounts of the world that they provide. In this sense, NRTs endeavor to act against what Dewsbury et al. (2002) call 'the reductive vampirism' inherent in trying to fix the world within particular structures, models, orders, and frameworks posited by the researcher. NRTs argue that there is always some excess here, something that escapes such framing, given that things keep moving. This means that the world is "more excessive that we can theorize" (Dewsbury *et al.* 2002: 437). This attention to process is also manifest in the modesty of those practicing research influenced by NRTs in terms of the claims they make about the world and in the way these accounts are produced and presented. Inspired by developments in the performing arts, NRTs often align themselves with a sort of methodological experimentalism that does not shy away from providing an open-ended account of the world (Dewsbury 2010a).

Subjectification

Thrift suggests that 'NRT' is "resolutely ... pre-individual. It trades in modes of perception which are not subject-based" (Thrift 2008: 7). Instead, NRTs are concerned with 'practices of subjectification'. 'Subject' here refers to our sense of self which we might assume is a constant accompaniment to our experiences of the world. Such selfhood is a bit like our identity, though not necessarily quite so easy to pin down into identifiable positions (i.e. those related to gender, race, ethnicity, age, religion, etc.). We might assume that our subjectivity is something that exists prior to the encounters we have and that it is through it that we make sense of those encounters. NRTs see things differently. For NRTs, subjects move from a secure and organizing position, present in advance of encounters, to become something that (provisionally and perpetually) arises from those encounters through these processes of subjectification. Sometimes, those processes unfold organically, but sometimes, they are also actively shaped; we attend school, are enculturated into specific social norms, and so on. Our sense of self, while potentially having some consistency, emerges and evolves over time here. This subjectification arises out of the world being "made up of all kinds of things brought into relation with one another by many and various spaces through a continuous and largely involuntary process of encounter" (Thrift 2008: 7). Subjectification proceeds from an ever-shifting composition of human and non-human things – various objects, people, technologies, texts, ideas, discourses, rules, norms, and so on – perpetually encounter and shape one another.

Embodiment

Such an attention to subjectification also leads NRTs to be interested in the human body and its co-evolution with things (Thrift 2008). Here, the body is not counted as separate from the world, but rather it is argued that the human body is as it is because of its

> unparalleled ability to co-evolve with things, taking them in and adding them to different parts of the biological body to produce

something which ... resemble[s] a constantly evolving distribution
of different hybrids with different reaches.

(Thrift 2008: 10)

By using 'hybrids' and 'co-evolving' here, Thrift is drawing attention to how our bodies, and their relations with the environments they are positioned in, are often tied up with non-human things. Anything from clothing, to contact lenses, to information communication technologies (ICTs), to cars and other forms of transport, all come to interact with our bodies. Such technologies mediate our relations with the world around us, impact on how we perceive those surroundings, and potentially augment what our bodies can do. Thermal clothing, for example, allows us to spend more time in cold environments than our bodies otherwise could cope with. In this, those cold temperature won't be felt as cold as they would without such clothing. Contact lenses bring the world into greater focus. Further, ICTs allow us to communicate over large distances and encounter images and sounds from a host of cultures that we otherwise might not directly experience. They broaden the horizons of our possible experience. And when it comes to transport, cars, for example, both vastly increase our ability to travel at speed but also lead to us developing different forms of spatial awareness – we (hopefully!) end up not just being able to perceive where the limits of our body are, but also the extremes of the vehicle as we navigate through traffic and other obstacles. As a result of such embodied relations with things, NRTs argue that "bodies and things are not easily separated terms" (Thrift 1996: 13). Therefore, NRTs aim to attend to the relatedness of the body and world and its constantly emergent capacities to act and interact.

Affect

This emphasis on the interconnected nature of our bodies with a whole host of non-human things is also closely connected to NRTs' desires to "get in touch with the full range of registers of thought by stressing affect and sensation" (Thrift 2008: 12). A key starting point for NRTs has been the realization that consciousness is in fact a narrow window of perception. At the outset, Thrift became fascinated by developments in neuroscience which cleaved a distinction between thought and action. In particular, experiments which identified a half-second delay between a body's action and the ability to account for that action showed that there was a whole lot going on in our bodies that we might not immediately be aware of or consciously in charge of (Thrift 1996). As a result, NRTs have called for more attention to be given to these pre-cognitive aspects of embodied life, these "rolling mass[es] of nerve volleys [which] prepare the body for action in such a way

10

that intentions or decisions are made before the conscious self is even aware of them" (Thrift 2008: 7). One way that this was articulated was through ideas of 'affect'. Affect does not refer to a personal feeling but rather to shifts in the state of our bodies which impact upon our capacities to act. These shifts are always going on but we're not necessarily aware of them. However, at times, these affects might come to be felt in our bodies. Think of how we might come to feel jittery after one too many cups of coffee or how we might come to feel down when we have one too many alcoholic drinks. Or it could be something less easily characterized like getting goosebumps or a shiver going down our spine when listening to a song. Such felt experiences emerge as something changes in our bodies – our heart rates shift, our central nervous systems become depressed, and sounds resonate in our bodies as well as in our ears. The outcome of that might be something that we can name as a recognizable emotion – we might say we feel excited or sad – but it also might remain something we struggle to put into words.

Agency

NRTs' interest in the body's co-evolution with non-human things and the affective relations that circulate amid that imply a certain understanding of the status of non-human things, be them animate or inanimate, human or animal, human-made or natural (though those distinctions may themselves be questioned by NRTs). NRTs ascribe a significant amount of agency to the non-human world. This comes through, for example, in references to 'heterogeneous networks' and the relations that make up such networks being 'flat' rather than based on a hierarchy where humans have the power to act and objects are acted upon. Rather than such distinctions between humans and animals or human and things, we are presented with reference to a range of actors or actants (or objects) that play a part in the functioning (or lack of functioning) of such networks. Those might be, for example, the social networks that support communities (though the designation 'social' itself already disguises a host of non-human actants in those networks) or it could be more obviously technical systems that are made up of a whole host of interactions across space and time (think of the internet and its various servers, cables, routers, computer terminals, technicians, webpages, signals, etc.). Ultimately, the emphasis in this is that humans are not the only ones who can make a difference in how life plays out.

The rest of the book

In this book, I aim to do two things. First, I aim to provide an accessible introduction to the key premises of NRTs. This is the primary aim of the text and is something that is often difficult to find within the existing literature on 'NRT'. This might appear modest, but the vast range of ideas, concepts, perspectives, and debates that have come to characterize NRTs make this more ambitious than it first sounds. Beyond that, though, my second aim is to articulate a particular version of what NRTs offer to geographic scholarship. That does not mean that I am going to proclaim what 'NRT' should be or do or how geographic scholarship more broadly should be done. There is enough of that in the existing NRTs-related literature. Instead, my aim here is to highlight a range of contributions NRTs make and take into account a range of critiques and developments that have emerged over the past two decades. At times, this will mean responding to those critiques, and in others, it will mean recognizing their merits and suggesting what might follow from them and/or how other geographers have constructively responded in their work. This will not add up to "the final word on 'NRT'" but rather, I hope, provide a starting point for readers to pursue their research in light of such ideas.

In doing this, the remainder of the book unfolds as follows.

The next chapter 'Nonrepresentational Theories and Geography' will build on this introduction by further outlining the emergence of 'NRT', focusing specifically on its geographic and intellectual lineages. This chapter will cover where and when (both in a geographic and disciplinary sense) NRTs came from, provide some further explanation of their key thematic interests, and elaborate further on why 'NRT' has proved to be both an instructive and problematic nomenclature. More specifically, the way that NRTs were overtly positioned in opposition to 'New Cultural Geography' will be explored. Connections to and differences from other earlier bodies of work in geography (Time Geography, Humanistic geography, and so on) will also be drawn out. This will show that NRTs did not necessarily present a complete 'break' in the history of geographic thought, but also that it has not meant 'business as usual' in a number of ways.

Chapter 1 is probably the 'heaviest' of the book in that it deals primarily with the ideas that underpin 'NRT'. From thereon in, though, each chapter is intended to act as a stand-alone introduction to a specific conjunction of 'non-representational theories and ...'. Each of these chapters takes an area/topic of geographic enquiry and explores the impact of NRTs on how it has been researched and understood. The order of these chapters is deliberate in terms of, broadly, a movement from key themes or concepts to more substantive areas of work. However, it should be possible to read them in any order or independently (and perhaps without the conceptual primer found in Chapter 1).

Chapter 2, 'Non-representational Theories and Practice', will explore further how practice has formed a fundamental starting point for NRTs. This emphasis on practice pervades the various core themes of NRTs and so much of the literature. This chapter provides a detailed consideration of the origin and nature of this focus on practice, both in terms of it being an analytical starting point – meaning that is has meant a focus on cognate terms such as performance, embodiment, and performativity – but also a direction for empirical inquiry for NRTs. More specifically, this chapter unfolds around a discussion of three key terms from within this work on practice: 'The Event', 'Rhythm', and 'Passivity'.

Chapter 3, 'Non-representational theories and Affect', will explore how affect has come to be a key concept within the NRTs literature as well as a source of some debate. Affect has featured in both conceptually driven writings, and it has formed a core concern for many empirically orientated studies. As such, Chapter 3 will provide a brief overview of some of the central concerns for such conceptualizations, including the complex relations of affect to connected concepts like feeling and emotion. Developing this, the chapter will explore three themes to show the range, nature, and scope of work developing and drawing on affect to think about a variety of geographic contexts and practices. This will focus on 'Animating the Everyday', 'Collective Affects', and 'Mediating Affect'.

In Chapter 4, 'Non-representational theories and Materiality', the focus will move on to how NRTs have rethought the way geographers understand and research materiality. This concern is by no means unique to NRTs. As such, the chapter will start by looking at the questions various geographers have asked about cultural geography's attention to the material world, particularly in terms of the emphasis of 'New Cultural Geography' on texts and representations. In looking to NRTs' specific contribution to these debates, the chapter will consider: 'What do we mean by "matter"?' and a host of different takes on how matter and 'materialization' might be understood.

In Chapter 5, the discussion will move onto what has been both a prominent and contentious conjunction between 'Non-representational theories and Landscape'. A key feature of the initial articulation of NRTs was the way they were contrasted with the work of 'New Cultural Geography'. This comes through most clearly in the particular ways in which NRTs have understood and approached landscape. Broadly, this sees a shift in emphasis from representations to practice and embodied experience. This shift will be further explored here, specifically through a discussion of how, arguably, the term landscape itself implies a fixity and finished character. In response, the chapter will look to the questions asked by NRTs about how landscapes come to be animated by people's practical engagement with them. In that, three themes will be explored in some detail, namely around ideas and practices of 'dwelling' in landscapes, landscape and mobility, and haunted/spiritual landscapes.

Chapter 6 moves on to discuss 'Non-representational theories and Performance'. This chapter will look at how creative and artistic practices have formed a key focus for NRTs since their inception in geography. In particular, the chapter will introduce the central idea that such performances are composed of generative relations between bodies and spaces. Here, performances are understood to shape the character of spaces but equally spaces (and their border social setting) are seen to 'act back' in the unfolding of a performance. This will be explored through two types of performances: dance and music. While not necessarily an obvious topic for geographers to study, dance constituted one of the first substantive areas of discussion for geographers developing NRTs. Further, NRTs have considered the performance and reception of music and sound, but in ways that depart somewhat from cultural geography's past concern with music as a sign of cultural diffusion or a source of identification at various scales of belonging.

The penultimate chapter will explore 'Non-representational theories and Methods'. Throughout the preceding chapters, a diverse realm of phenomena, encounters, relations, and entities will have been introduced into geographic scholarship. This, in turn, opens up questions around how scholars might 'do' research after NRTs. Therefore, this chapter provides an overview of various suggestions for the need to reinvigorate geography's research methods and methodologies in light of NRTs' arguments. In particular, the discussion will focus on geography's recent interest in various forms of image-based methods as a case study for such reinvigoration. That said, the chapter will also propose that readers should adopt a critical disposition toward such methodological proposals and propositions. Finally, the chapter will conclude by reflecting on the challenges of writing about research in light of NRTs.

The book closes with brief concluding chapter which flags key issues and agendas that, based on the preceding discussion, appear significant to NRTs' ongoing development and situation within geography and beyond.

Further reading

In this short dictionary entry, **Anderson (2009a)** gives an accessible introduction to NRTs that highlights where NRTs came from, suggests some elements of its focus which are related to other approaches to doing human geography, and addresses a few critiques that have emerged as a result. **Simpson (2017a)** provides another short introduction to NRTs. Again, some of the key themes found in NRTs are briefly introduced. This provides another initial starting point from which you could identify and pursue further reading from more research-orientated literature. **Patchett's (2010)** blog post provides a good starting point for further reading. Specifically, it both briefly introduces what NRTs focus on but also raises some critical points (specific around research methods). There is also some discussion of 'more-than representational' takes on NRTs. In this 'annotated bibliography', **Simpson (2015a)** provides both a range of short introductions to NRTs and the themes they have covered but also a list of references under each heading. Short annotations are included for each reference listed, explaining their focus. This may be useful in orientating you when it comes to what to read next (both now but also after each of the chapters in this book).

1

NON-REPRESENTATIONAL
THEORIES AND GEOGRAPHY

Introduction

This chapter focuses on the relationship between NRTs and geography. In particular, it will trace out some of the intellectual lineages of NRTs. The previous chapter introduced some of the background to NRTs emergence and identified some of their key thematic interests. This chapter will take this further by focusing in more detail on NRTs' 'origin stories', both in terms of the intellectual climate NRTs emerged within, and the literal geographies of their emergence and subsequent diffusion. As part of this, more will be said about the conceptual influences that underpin NRTs and the range of social theories and philosophies that have been introduced to geography. The chapter closes by reflecting on the different ways that NRTs have had an impact on the discipline of geography and the extent of that impact.

When it comes to the unfolding relationship between NRTs and geography, NRTs have been received, and continue to be received, in a range of ways. For many, NRTs have opened up a host of different and exciting questions about life's varied and varying geographies and have brought a new and diversified realm of matters into view. That said, a range of critical commentary has also been produced in relation to 'NRT'. This has included strong criticism of the basic points being argued, criticism of the particular conceptual position being taken, and, in that, disagreement with the critical comments Thrift made about other geographic approaches that NRTs sought to 'enliven'. These divergent dispositions noted, it is clear that NRTs have increasingly impacted upon the practice of research in geography in increasingly diverse ways. This will be shown here both in terms of the range of scholarship around NRTs that has emerged but also based on the broader diffusion of NRTs' arguments into what might be seen as 'mainstream' geographic work.

Origin stories

In their introduction to 'Taking-Place: Non-Representational Theories and Geography', Anderson and Harrison (2010a) suggest that there are a number of possible 'origin stories' for NRTs, none of which are in themselves determinate but many of which played some role in the development of NRTs in geography. These possible origins include:

> the on-going impact of post-structuralism on the discipline [of geography] and, in particular, the avenues for thought opened by the translation of the work of Deleuze and Latour; an emergent concern for 'everyday life' and the forms of embodied practice therein; a specific confluence of energies, research interests and institutional setting focused on the School of Geographical Sciences in Bristol in the UK throughout the 1990s; the gathering together and elaboration of non-representational theories by Nigel Thrift; the crystallisation of desires to find new ways of engaging space, landscape, the social, the cultural and the political; the influence of the UK's Research Assessment Exercise [now Research Excellence Framework] through which, in Human Geography at least, value was attached to single author papers and which promoted an academic climate wherein so called 'theoretical' interventions could be valued as highly as more 'empirical' studies; a simple generational shift between the New Cultural Geography and what would follow; an ever more extensive engagement by geographers with other social science and humanities disciplines; a cynical careerist fabulation.
>
> (Anderson and Harrison 2010a: 3)

In this section, I am going to pick out elements of such possible origin stories and explore them further. In particular, this section focuses on three of these. First, the relationship with and distancing from New Cultural Geography (and, with that, post-structuralism) will be discussed. This is perhaps the most commonly discussed and contentious of these origin stories. From there, I take a step back to think about some earlier engagements geographers made with other social sciences and the humanities, specifically through the developments of Time-Geography, Humanistic Geography, and concerns for 'structuration' as potential precursors in geography for NRTs. These literatures and relations tend to be less discussed when it comes to these origin stories. Finally, this section concludes with a discussion of the School of Geographical Sciences and the role Thrift (and others) played in working out from that specific space-time context.

New Cultural Geography and its 'dead geographies'

'New Cultural Geography' emerged during the 1980s as a prominent approach to geographic research. During 1980s, a range of (largely UK-based) cultural geographers became interested in the insights of the 'Cultural Turn' (see Box 1.1) and, with that, Cultural Studies (Jackson 1989) and the work of a range of 'Cultural Marxists' (Cosgrove 1984). These ideas were translated into a geographic mind-set, producing a very different cultural geography. The 'new' here refers to a distinction drawn with the well-established cultural geography of the 'Berkeley School', practiced primarily in North America in light of the pioneering work of Carl Sauer (see Sauer 2008 and Chapter 4).

In distinction to the 'Berkeley School', this New Cultural Geography argued itself to be:

> contemporary as well as historical (but always contextual and theoretically informed); social as well as spatial (but not confined exclusively to narrowly-defined landscape issues); urban as well as rural; and interested in the contingent nature of culture, in dominant ideologies and in forms of resistance to them. It would, moreover, assert the centrality of culture in human affairs.
>
> (Cosgrove and Jackson 1987: 95; though see Price and Lewis 1993)

Central to New Cultural Geography then was a specific attention to how culture is spatially constituted. This, in turn, entailed a focus on issues of power, ideology, discourse, and forms of representation enacted within space. Such concerns were perhaps most evidently manifest in work critically engaged with cultural landscapes – with the representation of landscapes and the people within them, with specific meaningful objects present in landscapes (statues, monuments, signage, and so on), and the practices through which landscapes were produced and maintained (see Wylie 2007a; Chapter 5). Over the years, such analyses were developed through engagements with, amongst others, semiotics (Duncan and Duncan 1992), iconography (Cosgrove and Daniels 1988), post-structural analyses of texts and representations (Duncan and Duncan 1988; Matless 1998), and feminist scholarship (Rose 1993).[1]

BOX 1.1 THE CULTURAL TURN

The 'Cultural Turn' refers to "A set of intellectual developments that led to issues of culture becoming central in human geography since the late 1980s" (Barnett 2009: 134). This is often associated with the emergence of New Cultural Geography. However, a shift toward an interest in culture also took place in a range of other sub-disciplines in geography, such as economic and political geography, given the suggestion that culture was increasingly important to economic processes and political conflict (Barnett 2009). The Cultural Turn is often characterized by a movement away from certain approaches and trends rather than a more positive determination – a move away from a certain form of Marxist analysis, a move away from positivism/quantitative approaches, and not being realist in orientation. In their place, we find interests in post-modernism and post-structuralism (as well as Cultural Marxism), qualitative and textual analysis, and a concern for how the world is socially constructed (Barnett 2004; see Chapter 2). A significant impetus behind this shift toward an interest in culture was the work emerging from the 'Birmingham Centre for Cultural Studies' and its "recognition of cultural diversity and processes of cultural change" (Scott 2004: 24). This has included concerns for various sub-cultures and 'low' cultural forms (as opposed to 'high' culture) (Jackson 1989) and issues of power and inequality (Barnett 2004). As such, the Cultural Turn leads to the argument that "Meanings are contextual, specific, and contingent. And this is where geography comes in: because of culture, *things happen differently in different places*" (Barnett 2004: 42).

In some of the early introductions to NRTs written by Thrift New Cultural Geography formed a key target of critique and something that NRTs were variously articulated against. However, New Cultural Geography was rarely named. This unfolded less through the direct citation of, or engagement with, examples from such New Cultural Geographers' work and more through the targeting of some of New Cultural Geography's conceptual influences. For example, Thrift (1996: 4) was critical of work informed by post-structural philosophers such as Michel Foucault and Jacques Derrida, in which, he claimed:

A hardly problematised sphere of representation is allowed to take precedence over lived experience and materiality, usually as a series of images or texts which a theorist contemplatively deconstructs, thus implicitly degrading practices.

This is not to say, though, that NRTs are not found in post-structuralist philosophies (Cresswell 2013a; Doel 2010; Wylie 2010; see Box 1.2). Rather, Thrift's target of critique here is a particular version of post-structuralism which led geographers to focus on, for example, the sorts cultural landscapes suggested above, taking them primarily (or even solely) as texts that can be read. The problem for Thrift is that these 'texts', whether literally a text or something treated 'as-text' (see Duncan 1990), form the primary focus of the analysis rather than the practices that they potentially play a part in shaping.

BOX 1.2 POST-STRUCTURALISM AND GEOGRAPHY

Post-structuralism is a name that is normally used to refer to a relatively diverse body of philosophical work that emerged during (and since) the 1960s in France (Harrison 2006). In terms of general characteristics, one important point for much of this work is that it is:

> profoundly suspicious of anything that tries to pass itself off as a simple statement of fact, of anything that claims to be true by virtue of being 'obvious', 'natural', or based upon 'common sense'. As a philosophy and a set of methods of doing research, post-structuralism ... exposes all such claims as contingent, provisional, subject to scrutiny and debate.
>
> (Wylie 2006a: 298)

From this, it is common to encounter claims that a whole host of things (values, meanings, identities, social structures, or hierarchies) are 'constructed' rather than true or essential and that such claims to truth need to be deconstructed through a process of radical questioning. Effectively, post-structuralism is suspicious of anything claiming to act as a foundation for thought or understanding. This suspicion meant, for example, "examining how social relations of power fix social practices, objects, events, and meanings as self-evident, given, natural, and enduring" (Dixon and Jones 2004: 80). Post-structuralism's role was to un-found such fixings.

Post-structuralism has been a significant influence for Cultural Geography since the 1980s. This included the growth in interest amongst New Cultural Geographers in representation – "the social mediation of the real world through ever-present processes of signification" (Dixon and Jones 2004: 87) – and questioning the politics and inequalities that are found in and are reproduced by these representations. However, given post-structuralism's internal diversity, post-structuralism has influenced other approaches to cultural geography. For example, certain strands of post-structuralism such as Actor Network Theory (ANT) have also influenced the development of more-than human geographies (see Whatmore 2002; Chapter 4). The development of NRTs in geography has equally been influenced by the work of ANT but also other post-structuralists, including Deleuze and Derrida (Wylie 2010; see Chapter 2).

In his chapter 'Steps to an ecology of place', Thrift (1999: 300) develops this critique of representationalism further in listing one of his theoretical 'dislikes' as work which takes the position that:

> human beings are engaged in building discursive worlds by actively constructing webs of significance which are laid out over a physical substrate. In other words, human beings are located in a terrain which appears as a set of phenomena to which representations must be affixed prior to any attempt at engagement.

Again, Thrift's target here is work where the focus falls upon the deliberate and deliberative production and attribution of meaning, in this case by subjects in relation to the environment they live within. It is not so much that Thrift is arguing that we do not do this. Obviously, at times, we will find ourselves in a situation where we quite consciously interpret and project meanings onto our surroundings based on our past experiences and current frame of mind. For example, we might reminisce with family members about how a particular scuff mark was made on a piece of now old furniture on the day you moved into a family home, you might feel wary walking down a street at night as you think about how it was the site of a friend being mugged recently or any number of other frequent or occasional occurrences. Rather, Thrift is arguing that focusing on the consciously thought about misses so much of what also goes on in our day-to-day interactions with the worlds we live in. This missing realm of worldly relation is perhaps most forcefully articulated in his paper with JD Dewsbury which critiques what they call 'dead geographies' and asks how such geographies might be brought to life (Thrift and Dewsbury 2000). Again, New Cultural Geography is not directly named here, but the target again is work concerned with "consciously planned codings and symbols" (Thrift and Dewsbury 2000: 415).

Cresswell (2012: 99) suggests that by such logics "Representation has been equated with a dead and already achieved work in stasis, while the lively, embodied world of the event [i.e. the concerns of NRTs] is one of generous and affirmative world in constant transformation". However, representations can and do fall within the world of relations and practices that NRTs seek to draw attention to. In this, NRTs call for representations to be understood in a particular way – as something that is generative of difference and not the re-presentation of the same, fixed, finished thing (Doel 2010). The 'enlivening' contribution of NRTs here, then, is argued to come in their focus on the performance of everyday life and the performative nature of this life, a focus on the tension between the creativity of the unthought everyday actions which we constantly undertake and their taking place within some kind of restrictive, normative situation. As Anderson and Harrison (2010a: 7) note:

> most of the time in most of our everyday lives there is a huge amount we do, a huge amount that we are involved in, that we don't think about and that, when asked about, we may struggle to explain. How did you know to come into the room through the door? How did you know that *that* was a seat?

Taking this further, so much of the meanings that do build up over time are not something we are actively involved in producing but rather accrue through such routine mundane practices. We develop attachments to specific places, we come to dislike others, we feel more or less comfortable in certain situations without necessarily ever *actively* thinking why that is, how it came to be, or how it might be changed.

Such positioning of NRTs and their contrast with the 'representational' work of New Cultural Geography can be questioned. First, Thrift is rather guilty of presenting something of a 'straw person' in his (often implied) critiques of New Cultural Geography. This has been a common, and often justified, critique of 'NRT'. In such critiques, it is argued that New Cultural Geography was far more nuanced in its accounts of, for example, cultural landscapes than it is given credit for by Thrift. As Cresswell (2012: 99) suggests:

> Geographers who wrote and write about representation (at least the good ones) were and are always trying to figure out how representation works in and with the world. The work of geographers informed by cultural Marxism, feminism, and poststructuralism was precisely about how meaning is unstable and unfixable, how power through representation is never complete, and how representation always works with practice and performance.

At the same time, though, it is also the case – as Cresswell (2012: 99) recognizes – that "an NRT take on representation has added a new liveliness to the way we think about representation as an act – a verb – rather than as a thing". There is something different about NRTs in the way that they, again despite appearances, critically engage with questions of representation and their role in the unfolding of practices. NRTs' focus on representation's performative, enactive nature does something (Anderson and Harrison 2010b; Dewsbury *et al.* 2002).

Second, others have argued against this sort of rupture-based understanding of the emergence of NRTs in relation to what went before. In some senses, we can see continuity *and* change here. As Wylie suggests (2010: 103):

non-representational approaches [can in fact demonstrate] ...the ongoing and evolving articulation of various branches of poststructural thinking and writing within Human Geography – an articulation that began, in geography, in the late 1980s and early 1990s [i.e. in New Cultural Geography], with the cultural turn's recognition of the crisis of representation, and its initial foregrounding of discursive and critical-deconstructive epistemologies.

If we think in this way, we lose the 'representational' versus 'non-representational' divide that Thrift's initial outlines set up. And, with that, we recognize on the one hand the diversity within such post-structural strands of thought and, on the other hand, the diversity of work present both within New Cultural Geography and associated with NRTs. Again, it is a question of evolving and developing frameworks of analysis that add in new questions, pose new problems, add points of nuance or disagreement, and raise different matters of concern rather than supplanting or moving on from what went before to what comes now.

Flashback: Time-Geography, Humanistic Geography, and 'structuration'

NRTs' relationships with New Cultural Geography present an obvious point of comparison given the way that NRTs emerged at a time when New Cultural Geography was both well established and prominent within geographic scholarship. However, there are other less commonly discussed or debated comparisons to be made – both in terms of difference and inheritance – with approaches in human geography that pre-date NRTs to varying degrees. This section will briefly pick up on three of these: Time-Geography, Humanistic Geography, and work concerned with 'structuration'.

Time-Geography

Time-Geography was developed by the Swedish Geographer Torsten Hagerstrand and others working with him at Lund University, Sweden, during the 1960s and 1970s. For Time-Geography, space *and* time (or space-time) were taken as resources that placed constraints on human activity. Time-Geography sought to develop a "respect for the conditions which space, time and the environment impose on what the individual can do" (Thrift 1977: 4). More specifically, Time-Geography attempted to "understand under what basic conditions linkages...[in space-time] develop and how such conditions can be changed in order to improve the quality of life" (Thrift 1977: 4). In this, time and space were understood in such a way that they existed 'outside' of the individual as objective and so measurable facets significant to human activity, especially movement. Individuals could be located in space and time, and such locatedness could be expressed through quantitative data. As Parkes and Thrift (1980) note, such spaces might include specific geographic territories, and such times might be understood as 'clock-time'.

Time-Geography's approached allowed individuals' movements in space-time to be mapped in various ways, perhaps most memorably in 'time-space budgets'. Here, the 'paths' of individuals across space and through time were charted, with the vertical axis representing time and the horizontal axis (or axes) space. As time unfolds, an individual's position in space is charted, generally starting at home and moving out to various locations. This can map the activities of a single individual or can include the mapping of multiple individuals who might come to coincide in space-time at various points (see Figure 1.1). A whole range of other modes of representation were developed from this that increasingly added detail and nuance into the diagrams. This included details of the specific locations visited or the mode of transport used, the representation of what constraints were imposed upon

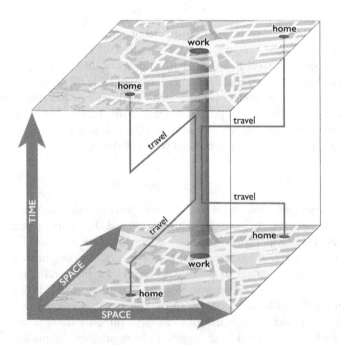

Figure 1.1 The space-time 'aquarium' (Author's own).

possible spatial range based on, for example, the activities undertaken, max-imum speed of travel, and so on.

While original in their attention to the level of the individual and in their introduction of time into geographic considerations (Parkes and Thrift 1980), Time-Geography was subject to a number of critiques. For example, Latham (2004: 124) notes how Time-Geography's time-space budgets pre-sented "little more than a ghosting of the individual's path through an envi-ronment". This then risks the production of "a mechanical, lifeless, picture of society" (Latham 2004: 124). Time-Geography's sense of space and time as 'outside' the individual risks missing something of the 'content' of human experiences of space and time. This can, for example, be contrasted to more 'behavioral' or 'humanistic' senses which would relate more to the *experience* of space and time. This might include our mental maps of the world – images of how places relate to each other based on our past movement between them – or to our personal sense of time passing – that could be time passing quickly when we are 'lost' in an activity or equally too slowly when we are engaged in something that doesn't interest us and want to be done with.

There are, though, a number of senses in which Time-Geography can be seen as both a background influence for the development of NRTs in geogra-phy and in how it could be considered something of a forerunner to them. First, Thrift explicitly acknowledges as much when he traces his own intel-lectual journey prior to his outlines of NRTs written in the mid-1990s. Thrift (1996) lists 'time-space' as one of four 'obsessions' and, in that, a concern for things like time-space budgets. Second, the specific nature of Thrift's own engagements with Time-Geography resonates with aspects of NRTs' eventual focus. Thrift, writing with Parkes, specifically notes a concern for combining Time-Geography's focus on 'outside' or 'locatable' notions of space and time with the more experiential senses discussed above (Parkes and Thrift 1980). There is an effort there to bring in a sense of what it is like to move through space and time, of the practice of movement as well as the how, where, when, and so on. It is not hard to see some of this in the work of NRTs when it comes to time, even if the terms through which it is discussed have evolved toward 'becoming', 'process', and 'difference' (Thrift 1999). Third, we find in a range of Thrift's discussions (and subsequent work) a concern for finding novel ways of documenting and 'diagramming' movement (see McCormack 2002, 2005). While these might be less 'scientific' in orientation than the rep-resentations produced by Time-Geography, they nonetheless aim at showing something of the movements NRTs have become interested in but which are hard to fully capture or put into words (see Thrift 2000; Thrift and Dewsbury 2000). Finally, and in a related vein, some subsequent work associated with NRTs has returned specifically to the space-time budgets of Time-Geography and has sought to develop them further in terms of better understanding the 'content' of space-time budgets and in finding innovative ways to expand upon their representation of movement in space-time through the inclusion of more qualitative/visual data (see Latham 2003a, 2004).

Humanistic Geography

Humanistic Geography formed a significant approach to human geography during the 1970s and 1980s. Humanistic Geography was a diverse body of work which drew on a range of different theoretical sources of inspiration, including phenomenology, Vidal de la Blache's 'geographie humaine', existentialism, social interactionism, and pragmatism (see Buttimer and Seamon 1980; Ley and Samuels 1978; Smith 1984). What that work held in common was a critical disposition toward 'spatial science' and quantitative approaches in human geography which had become prominent in the 1960s. More specifically, the emaciated version of the human that existed within spatial science – broadly, a rational actor whose decisions were motivated by economic concerns and full information – was felt to miss the importance of more humane facets of existence like beliefs, attachments, memories, morality, aesthetic judgments, and so on. Put differently, humanistic geographers "challenged what they viewed as an overemphasis on analytic simplicity that seemed to distance human geography from the creative and chaotic flux of everyday life" (Entrikin and Tepple 2006: 31).

'Creative and chaotic flux of everyday life'. Sounds familiar, doesn't it? You would be forgiven for thinking that the central concern of NRTs was being described here (see Cresswell 2012). Looking closer, other points of commonality emerge. Seamon's (1980) humanistic work on movement and dwelling – and particularly his articulation of the 'body-ballets' that sustain so much of our (unthought) actions in accomplishing various day-to-day tasks – drew heavily on the phenomenology of Merleau-Ponty. That phenomenology also formed a prominent initial point of inspiration for Thrift when it came to outlining his concern for embodiment and the prosthetic nature of the human body. Further, Seamon's focus in that on "body-subjects that respond contextually and preconsciously through their sensible actions, rather than mental subjects who cognitively interpret space and then proceed to act in ways that are based on these cognitive interpretations" (Ash and Simpson 2016: 51) could very well be a description of a range of work associated with NRTs (see Chapter 5).

Based on this, it is surprising that Thrift and others rarely (if ever) cited humanistic geographers such as Seamon in the earlier articulation of NRTs (though more recently, see Ash and Simpson 2016; McCormack 2017). There are a number of possible explanations (if not quite justifications) for this. Merleau-Ponty is a common point of influence here for each body of work. However, that influence has unfolded along different lines given the different work Merleau-Ponty has been combined with. For example, a host of more overly post-humanist or anti-humanist theories have been significant for NRTs and so have pushed discussions there in different directions (see Wylie 2006b). Further, a commonly critiqued central concern of Humanistic Geography was related to its aspiration "to describe and explain the human experience of nature, space and time" (Buttimer 1976: 278). This came under critique for the way that it suggested a singular 'human experience' of space and so that various aspects of socially influenced difference in this experience were not recognized (for example, see Rose 1993). While 'NRT' has been accused of a form of universalism in its accounts (see Chapter 3), a key focus amongst NRTs has been on difference rather than sameness. This is a key point that will recur in various chapters here. The emphasis throughout NRTs on becoming, change, and differing in the unfolding of day-to-day life pushes in quite a different direction than the aspirations of Humanistic Geography as described by Buttimer. Nonetheless, as with New Cultural Geography, the balance between evolution and break when it comes to the trajectory of geographic thought from Humanistic Geography to NRTs can be questioned.

'Structuration'

'Structuration' refers to a theory developed in the 1970s and 1980s by the sociologist Anthony Giddens, though it has also been associated with the work of others (see Thrift 1996). This theory sought to interpret the intersections of human subjects and the social structures that they were involved within (Gregory 2009). Structuration was developed in an attempt to reconcile a common problem of social theory in that many discussions of social life tended to either prioritize human agency or prioritize the structures of that society. The former focused on the voluntarism of the intentions and actions of subjects and was often associated with Humanistic Geography. The latter focused on the determinism of the specific political-economic system of a society and the limitations that imposed and was often associated with Marxist approaches (Thrift 1996). Giddens sought to give neither the human agent nor the social structure priority (Dyck and Kearns 2006). Instead:

> Giddens proposed to treat the production and reproduction of social life as an ongoing process of *structuration*. In this view, 'structure' is implicated in every moment of social interaction – 'structures' are not only constraints but also the very conditions of social action – and, conversely, structure is an 'absent' order of differences, 'present' only in the moments of social inter-action through which it is itself reproduced or transformed.
>
> (Gregory 2009: 726)

In thinking about this structuration, Giddens proposed a range of concepts to help understand the dialectical interactions of structure and agency (see Gregory 2009). This included 'reflexivity' which refers to how the (re)production of social life is an accomplishment of skillful subjects who use (and transform) rules and resources. However, this also recognized the presence of more 'practical' action based on "tacit knowledge that is skillfully applied in the enactment of courses of conduct, but which the actor is not able to formulate discursively" (Giddens cited in Thrift 1996: 71). Further, the 'duality' of social structures is recognized in terms of their 'recursiveness'. Here, structures are taken both to be the medium and output of the social practices that act to (re)produce social systems. As Thrift (1996: 69) explains: "individuals draw upon social structure. But at each moment they do this they must also reconstitute that structure through the production or reproduction of the conditions of production and reproduction".

Given the centrality of social practices to NRTs, it is not hard again to see the import of structuration here. Again, this is specifically signaled by Thrift (see Thrift 1996). However, and again, it is the case that the language through which this is talked about has changed quite radically. While debates based on the language of structure-agency might have become, in Thrift's (1996: 2) words, "passé" (though see Dyck and Kearns 2006), that is not to say that a concern for agency, and increasingly, more-than human agency, does not remain a key focus of NRTs. Equally, the context of social practices – the specific, situated nature of practices in space and time – remains a central, if debated, concern for NRTs (Thrift 1996; also see Chapters 2, 3, and 6).

The 'Bristol school' of cultural geography?

> One day an account will be written of the role of Bristol in this particular episode of the history of geographic thought.
>
> (Cresswell 2012: 96)

Another way that we can think about NRTs' origin stories is more literally geographic in focus. NRTs have come to be closely associated with the School of Geographical Sciences at the University of Bristol (see Anderson and Harrison 2010a; Cresswell 2012; Lorimer 2015) (see Figure 1.2). In many ways, the scene that the Introduction to this book opened with could have been seen from the postgraduate offices at the School of Geographical Sciences ('Browns'), overlooking the intersection of Queen's Road, University Road, and 'The Triangle' outside them.

Figure 1.2   The School of Geographical Science, University of Bristol (*Alamy*).

There are a number of reasons for this association. First, Nigel Thrift – the geographer first and most commonly associated with 'NRT' – was based at the University of Bristol from 1987 until 2003 when the majority of his key papers articulating what NRTs are were written (see Thrift 1996, 1997, 2000). Second, Thrift was amongst a number of scholars there at this time interested in a range of conceptual insights from social theory and continental philosophy that challenged a range of geography's habitual ways of thinking at the time. This included, for example, the work of Sarah Whatmore (at Bristol from 1989 until 2001) on 'Hybrid Geographies' (see Whatmore 2002). Third, Thrift was involved in the supervision of a range of graduate students at Bristol around the time those introductions to NRTs were written. A number of these graduate students in turn engaged with the theories that underpin NRTs and have subsequently gone on to carry out research that is either influenced by NRTs or has specifically sought to further their development both at Bristol and beyond (for example, Dewsbury 2000; Harrison 2000; McCormack 2002). Finally, something of a legacy has been left in the School of Geographical Sciences at Bristol both in terms of (until recently) the continued presence there of Thrift's past postgraduate students (and/or their postgraduate students), but also through, for example, the MSc in Society and Space course (see Box 1.3).[2]

BOX 1.3 MSC IN SOCIETY AND SPACE AT THE SCHOOL OF GEOGRAPHICAL SCIENCES, UNIVERSITY OF BRISTOL

The MSc in Society and Space at the School of Geographical Sciences at the University of Bristol has played a significant role in the development of NRTs. The MSc began under the direction of Nigel Thrift in 1992 and "aimed to provide a thorough understanding of the theoretical debates around issues of society and space and how these translate into practical research agendas and the formation of policy" (Haggett *et al.* 2009: 24). The course takes the form of a one-year program of study based around a series of modules (mixing theory, methods, and 'topics') and a dissertation. From the start, students on the program have been introduced to a range of social theory and philosophy that held a significant influence on the initial development of NRTs. More recently, this theoretical focus has evolved based on NRTs' continued development (and staffing changes that have taken place within the School).[3] In addition to those formally registered on the program, initially, it was common for some graduate students to 'sit-in' on taught modules during their first year and, rather than complete the dissertation component, carry on with doctoral research directly. However, given the research training requirements of the Economic and Social Research Council, this program now commonly acts as the first step toward subsequent doctoral study.

It is possible to see both the theoretical influences taught in this program and the increasing development of NRT-related work from the dissertations completed as part of this program.[4] During the period 1993–2013, there is a quite notable shift in the focus of the dissertations completed here. While there are exceptions, during the mid-late-1990s, the influence of New Cultural Geography is very evident in the dissertations focused on social and cultural themes. A number of dissertation titles here include reference to specific cultural landscapes, representation, and/or 'the representation of' specific places or peoples. Equally, their reference lists include reference to the work of a host of well-known New Cultural Geographers and the sorts of post-structural and Cultural-Marxist influences they drew upon. With some contrast (and again with some exceptions), looking to the early years of the 21st century, we find dissertation titles referring to performance, performativity, practice, encounters, and so on and references to a host of non-representational thought (Deleuze, Latour, Merleau-Ponty, and so on). And that trend increases in those social and cultural dissertations as we move toward the present.

In some ways, this specific geographic focal point for NRTs' emergence into geography could be presented as something that undermines their significance. Locating 'NRT' within a single department at a single University, and basing that around one prominent academic based there, could be used to suggest that 'NRT' occupies a 'niche' situation in geographic scholarship (see Nayak and Jeffrey 2011). However, NRTs would not be the first significant 'School' of cultural geography to develop or be discussed in such a way. Although the timeframes are radically different, analogies could be drawn to the Berkeley School of Cultural Geography and its emergence in/diffusion throughout the United States.

The Berkeley School of Cultural Geography emerged from the University of California, Berkeley in the early part of the 20th century. Central to this was Carl O. Sauer's writings which "served as the main impetus for North American geography's move away from environmental determinism toward more cultural and historical studies of human-environment relations" (Mathewson 2011: 58–59). Broadly, for environmental determinism, the "environment modifies the physique of a people... By imposing upon them certain dominant activities" (Semple 1911: 35). In contrast, for the Berkeley School, "The cultural landscape is fashioned from a natural landscape by a culture group. Culture is the agent, the natural area the medium, the cultural landscape the result" (Sauer 2008: 103). The Berkeley School approach to cultural geography was so prominent that it was suggested that "cultural geography in [the United States] was Carl Sauer" (Wallach 1999: 130). While Sauer and the Berkeley School are generally thought of as synonymous, there was very much the development of a School here. By the end of his working life, Sauer had overseen 37 doctoral dissertations. With a handful of exceptions, this group formed the core of the Berkeley School (Mathewson 2011). Furthermore, by the 1960s, a number of Sauer's past doctoral students were producing their own doctoral students. By the end of the century, this added up to five generation's worth of students totaling nearly 200 academics distributed across some 20 Universities (see Brown and Mathewson 1999). And that is not forgetting those less easily placed within the Berkeley family tree such as those who came to hold an affinity for such work without necessarily having been formally associated with it.

For NRTs, we can see *signs* of something similar, though on different scales. 'NRT' (in the singular) has been strongly associated with Thrift, with the two almost becoming synonymous. At the same time, though, NRTs were a relatively late development in Thrift's academic career. While in retrospect a trajectory toward NRTs can be seen, much of his work during the first half of his career was not directly tied up with NRTs. Equally, many of his earlier PhD students have not gone on to work on this area.

That said, from the mid-1990s, a number of Thrift's PhD students did come to develop research now strongly associated with NRTs. Subsequently, those PhDs have gone on to pursue research related to NRTs in a number of other Universities as well as at Bristol. That is not to say that they have adhered strictly to Thrift's 'take' on NRTs; NRTs are diverse in focus and approach, and it isn't always easy to mark out their boundaries clearly or definitively. As such, these individuals have continued to develop and expand work relevant to NRTs or broadly associated with them beyond the point where Thrift's own career veered more toward 'senior management' and so away from academic concerns. In turn, a second generation of graduate students have emerged from this, again spreading further both in terms of geographic reach and research focus. This second generation (of which I am a part) has in turn further elaborated upon the focus of NRTs both in terms of empirical focus and conceptual frame of reference and has commenced work with a potential further generation of graduate students. This has meant, at times, moving from the language of non-representation toward concerns for 'new materialisms' or 'post-phenomenologies', amongst others. That said, an inheritance can be seen here in terms of the background focuses within this on practice, process, embodiment, more-than human agency, and so on.

Throughout the above, there are of course a number of others who are not directly affiliated with Bristol and/or Thrift but who are commonly associated with NRTs. There are a number of ways in which this can be broken down. There are those who have overtly engaged in the development of NRTs. There are also those who have acted as sympathetic or critical commentators. There are those who have conducted work that might not be immediately associated with NRTs but are moving in various parallel directions to it (often on work on more-than human or relational geographies). These individuals can be variously identified from published commentaries, research publications, and their inclusion in conference sessions and edited collections specifically on NRTs. And again, there are multiple generations of scholars here with both established academics and their doctoral students, many of whom have taken up academic careers and have conducted work related to NRTs. Collectively, this adds in a range of universities, both in the UK and beyond to 'NRT's' extended family tree as well as a range of different specific empirical and conceptual questions. And this is all before getting to those who might have been variously influenced by the arguments of NRTs but do not necessarily (overtly) present their work as such. Extensive work emerging in geography over the past decade or so on 'affect', for example, attests to this (see Chapter 3).

This relatively length elaboration is not intended to suggest that NRTs will reach the extent that the Berkeley School did in terms of its dominant influence, the number of scholars working in this area, or in terms of number of generations that this unfolds across. Rather, it is simply to question some relatively common suggestions that NRTs do not extend significantly beyond the work of Thrift and a select few often associated with him. There is something of a Bristol School here, but it is not one that is strictly limited to that single site.

Conceptual influences

Having thus far considered the relationship between geography and NRTs in terms of other conceptual approaches and the geography of their development/diffusion, in this section, I look in the opposite direction. Rather than looking inward at NRTs' relationships with geography, this section looks outward to consider what theoretical influences NRTs have been drawn from in an effort influence how geography is done.

From the start, NRTs have been eclectic in their conceptual origins (see Cadman 2009). The thing that held such thinking together was a concern the "taken-for-granted which we can never take for granted" (Thrift 1999: 302). A range of such work can be seen in Figure 1.3. Such theoretical interests continue to develop and expand. Therefore, rather than focus on specific theorists, this section will focus on some of the central schools or bodies of work referenced in Thrift's earlier outlines. The key points of reference briefly discussed here are: Practice Theory, Phenomenology, ANT, Vitalism, and Performance Studies.

Practice Theory

Thrift (1997) suggested 'the theory of practice' as an alternative name for the style of work he was trying to outline. Connected to that, inspiration has variously been drawn from what could be called 'practice theory'. This covers a range of work including: North American Pragmatism (see McCormack 2013), Ethnomethodology (see Laurier 2010), and a range of Continental Theorists such as Walter Benjamin, Pierre Bourdieu, and Michel de Certeau (see Thrift 1996, 1997). While differences exist amongst such work, NRTs circulate here around their concern for "mundane everyday practices ... that shape the conduct of human beings towards others and themselves in particular sites" (Thrift 1997: 127). Here, there is a significant realm of practical expertise that is often enacted without conscious consideration and so does not ever come to be represented or accounted for by such practitioners. In such unthought practice, we are often conforming to what is 'normal', what is expected of us, what we always do, and what others also do. A central motif that has emerged throughout such discussions is that of 'performativity'. As Dewsbury (2000: 475) notes

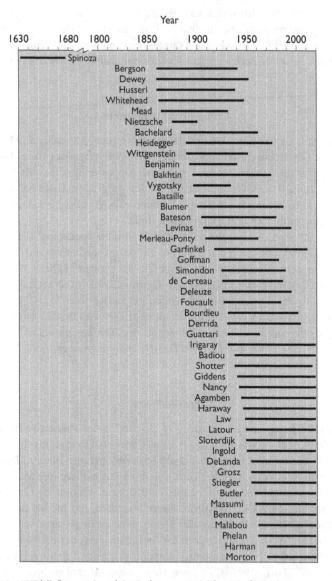

Figure 1.3 NRTs' 'life-time-lines' (*Author's own, expanded from Thrift 1999*).

Performativity is a slippery term indubitably linked to the idea of a performance, but, regardless of the multiple instances by which a performance might come to be defined ... the performative slips across, beyond, and through such actual renditions. In this sense, whilst constituting a discrete act – the performance – the performative is not itself a concept signifying such an act.

The work of Judith Butler in particular has underpinned NRTs' attention to how subjectivities are constantly being negotiated through a relative balance between unthought and ongoing practices of norm imposition/recitation and the occasional transgression that might take place within that (Thrift and Dewsbury 2000). That said, such understandings of performativity have also been spliced with theories which might be understood to be less 'austere' and more open to difference, change, playfulness, and ultimately forms of agency (see Dewsbury 2000). Such practice theories (and related terms like performativity) are explored in much greater detail in Chapter 2.

Phenomenology

It was mentioned earlier that phenomenology formed a key point of interest for Humanistic Geographers and a source of inspiration in the development of NRTs. There are two general directions in phenomenology that were picked up by Thrift in his initial engagement with non-representational thinking: a focus on embodiment and a concern for 'dwelling'. The former is most commonly associated with the writings of Maurice Merleau-Ponty. In this, Merleau-Ponty emphasized that we are and can only exist in the world as bodies. All of our 'lived experiences' are embodied experiences. As such, any practical action that we engage in is, by implication, embodied action. With that, we start to realize the importance of both our tactile engagements with the world and the various 'stuff' that exists within it. These embodied engagements are 'sensuous' – our bodies feel space as much as occupy it. In this, we blur any distinction between our body and the objects we encounter (Thrift 1996). Moving to the latter, we encounter the writing of Martin Heidegger. Here again, we find an emphasis on the 'world disclosing' nature of variously skillful practices. Heidegger's work draws a distinction between the sort of abstract, detached mental planning that human beings might distinguish themselves in being capable of doing – for example, imagining and planning out the structure of an essay or the location of a range of photos in a montage on a wall – and the involved activities that we engage in in specific environmental contexts. Here, it is the latter that provide the grounds for the former to take place, not the other way around. It is through practical action that the world comes to make sense to us rather than prior mental imaginings (Thrift 1999).

NRTs' relationship with phenomenological ideas perhaps comes through most strongly in work concerned with landscape. These engagements, while starting with Merleau-Ponty and Heidegger, have expanded to other phenomenologists and have been quite critical of phenomenology in light of developments in post-structuralism and vitalist ideas (see below), amongst others (see Ash and Simpson 2016). NRTs' relationship with phenomenological ideas will be returned to in some detail in Chapter 5.

ANT

Actor-Network Theory (ANT) has played a very significant role in the development of NRTs. This can be seen in shared interests in 'technologies of being' and the 'hybrid' nature of human's inhabitation of space, the relations of humans to non-human objects or things (Thrift 1996, 1997). ANT is often associated with the writings of Bruno Latour and emerged during the 1980s as a response to certain key trends in the study of society. In particular, rather than 'the social' being understood as an explanatory tool for ANT, 'the social' itself was taken as something to be explained (Latour 2005). Those developing NRTs were drawn to ANT as part of an effort to understand how individuals' practices might come to play a part in process of social formation (Thrift 1996). In this, ANT developed a range of conceptual tools. Notably, this includes the language of 'actors' or 'actants' rather than talking of subjects and object. For ANT, these actors can be human or non-human, animate or inanimate; there is a 'symmetry' or 'flatness' here for ANT between humans and pretty much everything else. This is not to say that everything has the same agency, but rather that decision should not be made in advance of any given relation over what agency any given entity should or does have. Agency doesn't exist in advance but rather emerges from relations. Further, these actors can exist at multiple scales – bacteria, an individual human, a cat, a crowd, a multi-national corporation, and so on can all be understood as actors here. It is from the activity of these actors that 'the social' is both constructed and constantly maintained. Society here is constituted by 'heterogeneous networks of association' that are more or less durable and are variously sustained, fall apart, or evolve depending on the interactions that take place within them.

The arguments of ANT have not been embraced by NRTs in their entirety. This can be seen perhaps most clearly in terms of the 'flatness' it seeks to impose on the world (see Rose and Wylie 2006). However, ANT has provided a significant starting point from which to consider a whole range of points around human relations with technology and materiality more broadly. ANT will be discussed in more detail in Chapter 4 when it comes to NRTs' engagement with understandings of materiality.

Vitalism

In addition to phenomenology, a range of other continental theorists who could be collectively referred to as 'Vitalists' have been significant to the development of NRTs in geography and further contribute to the development of something of a 'flat' account of agency such as that developed through ANT. This body of work has a lengthy and varied lineage, including philosophers such as Henri Bergson and Friedrich Nietzsche. The most prominent source of influence here though has been the work of Gilles Deleuze (and his writings with Felix Guattari). As Cadman (2009: 2) notes

> The philosophy of Deleuze and Guattari in particular has offered tools for those geographers who want to escape phenomenology's largely human-centered understanding of (embodied) practice and connect with the impersonal and transversal forces of the world.
>
> (Also see Roberts 2019a)

Rather than starting from a focus on embodied (i.e. human) experience, such Vitalist work starts from the perspective of a more general concern for 'life' and considers how phenomena like embodied experience emerge from the midst of a ceaseless variation that unfolds from the vast array of encounters that constitute day-to-day living. Again, these encounters are not just about humans but include various more-than human entities and objects. Rather than being discussed in terms of 'actors', they are more commonly referred to as 'bodies' here. Again, 'body' takes in much more than we might normally associate with it – an idea, for example, would be referred to as a body of thought. Within that, concepts like 'the event', 'assemblage', and 'affect' emphasize the relational composition of just about everything. Here, relations take place across a wide variety of time scales and spatial distances and proximities, at different degrees of intensity and across a range of shorter or longer durations, and with a host of human and non-human participants. In this, a range of forces interact bringing about a range of happy or unhappy encounters which either expand or limit what various 'bodies' are capable of (Thrift 1996).

Vitalist philosophies and the work inspired by them often employ quite a specific conceptual vocabulary. This makes this one of the less accessible sides of NRTs. It also offers some of the most radically different conceptual frameworks when compared with those that came before in geography and which might be understood as a more general 'common sense' understanding of the world. As such, these ideas are returned to and explored further in a number of chapters in this book. This includes the next chapter in terms of its discussion of 'the event' but also in Chapters 3 and 4 in their respective discussions of Affect and Materiality.

Performance Studies

Performance has been both a thematic and empirical focus when it comes to NRTs. This mirrors in some ways the central debates that NRTs have drawn upon when it comes to the Performance Studies literature. While diverse in focus and concerns, NRTs have picked up on a distinction made therein between something that 'is' performance and something that is taken 'as' performance. While not necessarily a neat binary, this allows for a range of themes to emerge that complement the focus of the various conceptual influences discussed so far.

In the sense of 'is' performance, we have an attention to various forms of theatre and musical performance as well as performing arts like dance. Again, these have formed the empirical topic of study for a range of work developing NRTs. Equally, a number of conceptual themes of interest to geographers working with NRTs have been drawn out from the Performance Studies literature on such arts (see Thrift 2000). This includes a concern for how it is that performances can produce, if temporarily, specific space-times of experience. For example, it has been shown that these space-times can be 'liminal' and so call existing social orders into question and allow spaces of possibility to be opened up. In this context, Thrift (2000) draws attention to the sorts of radical street theatre that proliferated in the 1970s. Equally, such performances are also seen to potentially be part of the performative reiteration of societal norms and that such reiteration can extend some way beyond the specific space in which performances take place and into the domain of the everyday. Certain genres of dance, for example, might reiterate gender norms when it comes to the respective roles of those involved.

Emerging from this engagement with performances, there is also the question of how such performances might be researched, recorded, and represented or re-presented given their temporally specific occurrence. There is very much a question of their non-representable nature. This concern for issues of 'liveness', though, has had multiple positive outcomes. It has suggested a need for methodological experimentation in how we might attend to phenomena that are fleeting and resist being fixed in space and time, and it has led to a recognition that various forms of recording and representation themselves do something. While something might be lost when a performance is filmed and replayed, the replaying itself has its own powers in the mediation it performs and the space-times of experience that produces. There are then a whole host of question to be asked of such mediated geographies. Chapter 6 will look more closely at this interest in performance in the context of 'is' performance, and Chapter 7 will return to the issue of liveness in the context of the sorts of methodologies and writing strategies that have been developed in light of NRTs.

Thinking in terms of 'as' performance, we start to recognize that various aspects of everyday life involve the performance of certain roles. This again echoes discussions from practice theory and interests in performativity discussed above as well as bodies of work such as 'social interactionism' that looks in close detail at how people interact in a range of everyday settings. Here, the everyday becomes a stage where different roles are performed – knowingly or not – depending on the space that is occupied, who is present (or absent), what codes of conduct are present (and adhered to), and the specific competencies of those involved (Thrift 1996). A concern for the performance of everyday life – 'as' performance – permeates throughout this book but is most prominent in Chapter 2's discussion of practice.

Conclusion

...before this excitable bandwagon pulls away we would like to momentarily pull the hand-brake...These ideas, although currently fashionable, remain largely the preserve of a select band of cultural geographers. Within much geographic research this is still something of a cult pursuit rather than mainstream practice. In our own institution and many others NRT is as yet not widely taught across mainstream geography curricula. We would agree that presently NRT is creating something of a stir, but this may turn out to be no more than an eddy in the backwater of cultural geography. Whether these ideas have reached their high-water mark, or will endure beyond the next decade – in the way that Marxist, feminist, post-modern and post-colonial approaches have succeeded – still remains to be seen.

(Nayak and Jeffrey 2011: 291)

It might seem odd to end a chapter on 'NRTs and Geography', a chapter near the start of a whole book devoted to 'NRT', with a quotation that both calls for a tempering of enthusiasms for the topic being discussed and that suggests the relatively limited significance of the body of work being considered. However, there are a number of reasons why it is worth considering this quote in some detail. In part, this is because the sentiments of this quote have appeared a number of times over the years of 'NRT''s development. Further, it also raises important questions about NRTs' relationships with, and impact upon, contemporary geography. It also asks questions about NRTs' geographies, their reach, and their impacts upon geographic scholarship.

First, as has been shown throughout this chapter, NRTs' relationship with past geographic thought is more complicated that might have first appeared to be the case based on Thrift's introductions to NRTs. NRTs did not necessarily present a complete 'break' in the history of geographic thought. NRTs equally have not meant 'business as usual' in a number of ways. With that, the degree to which NRTs' key conceptual references are a temporarily 'fashionable' pursuit can be questioned. In fact, it has been shown here that some of this interest is not part of a specific fashion trend but is actually an extension of a range of long-held interests amongst cultural geographers, stretching back 30 or 40 years. For example, there are both points of resonance and disagreement with work in Humanistic Geography, specifically around phenomenological theories and their usefulness in understanding the practice of everyday life. The contemporary usefulness of phenomenology for geographic scholarship continues to be questioned and developed (compare Ash and Simpson 2016; McCormack 2017; Roberts 2019a; Simonsen 2013; Wylie 2010). Equally, NRTs find both commonalities and differences with New Cultural Geography in their respective engagements with post-structuralist ideas. While this can lead to a 'texty' focus on 'social construction' through various representations and discourses, it also opens ways for thinking about more 'radical' forms of construction through affects and more-than human encounters. The specific points of reference in such work may be a little different – less Foucault and more Deleuze – but that doesn't mean that there are not common themes and concern over relationality, 'construction', and so on. And, there are also common referents, even if different aspects of their work are emphasized (for example, the Foucault of 'Discipline and Punish' for New Cultural Geography versus the Foucault of 'The Birth of Biopolitics' for NRTs (see Cadman 2009)).

Second, in terms of NRTs' supposed 'backwater' status, again, this can be questioned. While a tag like 'Bristol School' might imply that NRTs occupy some kind of southwestern backwater within UK higher education, this would be to ignore the ongoing diffusion of non-representational scholarship around UK higher education and beyond. This can be seen in a wide range of research-related outputs in academic journals. Equally, NRTs are increasingly present in a range of undergraduate textbooks on human geography and cultural geography, allowing for its permeation in various forms into undergraduate curricula. While that could be argued to be a product of the prolific output of a select few, there are other signs of a tipping point here. As Cresswell (2012: 96) suggests, "Edited collections like *Taking Place* are one sign of a particular set of ideas or theoretical approaches reaching a critical mass…an event that signaled the maturity of a set of theoretical approaches in human geography". Finally, we now have more specifically focused edited texts like Vannini's (2015a) collection on 'Non-representational Methodologies' and a number of sole-authored monographs (see Anderson 2014; Andrews 2018; McCormack 2013).

Equally, there is also the broader take-up of NRTs by scholars less directly associated with Thrift. That might relate to those who've fully climbed on board the NRTs 'bandwagon', but equally, it draws in a range of broader scholarship which now commonly talks in terms of practice, embodiment, affect, and the like. There is a sense that a host of 'more-than representational' scholarship has emerged whereby the insights of NRTs act as a 'modest supliment' to a range of geographic approaches, including those more associated with New Cultural Geography (Lorimer 2008). Rather than representational geography versus 'NRT', we can see a range of work employing a 'both and also' approach (Anderson 2010). We only have to look to the programs of major geography conferences like the RGS–IBG Annual Conference and the Association of American Geographers Annual Meeting to see a proliferation of sessions focused on such topics, something that filters through into the pages of a range of geography journals.

Further reading

Cadman (2009) provides a brief introduction to NRTs in terms of their thematic points of focus and conceptual influences. In particular, this looks at a range of philosophies that have influenced NRTs and a number of bodies of work – including ANT and Performance Studies – that NRTs have drawn upon. There is also some discussion of the implications of this and how NRTs have been received by geographers. **Anderson and Harrison (2010a)** provide a detailed overview of both the emergence of NRTs and their conceptual concerns and influences. This includes attention to the ways in which NRTs evolved and changed over more than a decade of work – specifically in terms of a move from an emphasis on 'practice' to a concern for 'life' – given the relative prominence of the different conceptual influences discussed in this chapter. This is a relatively advanced-level introduction but it is very accessibly written. **Cresswell (2012)** provides a critical review of Anderson and Harrison's collection 'Taking-Place'. Cresswell presents himself as an 'interested skeptic' here. In this, a range of points about NRTs come under question. This is useful in focusing on NRTs' relationship with a range of prior work in geography, including Humanistic Geography and New Cultural Geography. **Vannini (2015b)** provides an overview of NRTs' main thematic interests with a specific focus on how this translates into conducting research. This touches upon the influence of a range of the conceptual reference points for NRTs discussed in this chapter and develops an account of their implications for what research might look like in light of them. This is argued to add up to a specific style of working that is concerned with events, relations, practices and performances, affects, and backgrounds.

Notes

1 There is not sufficient space here to do justice to the diversity of work present in New Cultural Geography or the ways in which this evolved and diversified from the 1980s until the present. For various takes on this, see Chapters 1, 5, and 6 in Horton and Kraftl (2014), Chapter 3 in Anderson (2010), or Chapter 2 in Mitchell (2000).
2 See http://www.bris.ac.uk/geography/news/2017/25-years-of-society--space.html.
3 See https://bristolsocietyandspace.com/ for an indication of current themes covered.
4 This claim is based on the analysis of both the titles and reference lists of dissertations from this course held in the Wills Memorial Library at the University of Bristol.

2

NON-REPRESENTATIONAL
THEORIES AND PRACTICE

Introduction

Practice has formed a fundamental starting point in the development of NRTs. This emphasis on practice pervades the various core themes of NRTs and so much of the associated literature. This chapter provides a detailed consideration of the origin and nature of this focus on practice, both in terms of it being an analytical starting point – meaning a focus on cognate terms such as performance, embodiment, and performativity – but also a direction for empirical inquiry in light of NRTs. Such concerns have ranged from things as apparently mundane as going for a coffee (Latham 2003a; Laurier and Philo 2006a, 2006b); to going for a walk, getting the train, or going for a drive (Bissell 2009a; Thrift 2004a; Wylie 2005); to listening to music at home (Anderson 2005) or at a pub (Morton 2005).

Practice as a conceptual lens connects a diverse set of concerns common to NRTs around the functioning of social situations, situations constituted by a vast array of activities. Returning to the opening scene from this book's introduction, that space-time on the edge of a University campus included anything from the banal (walking, crossing a road, talking on a phone, speaking to a friend, sitting on a bench, and so on), to the creative (the production of the music being listened to on headphones or just thinking about the day to come and what that individual might do), to the contentious (restricting the rights of certain individuals to be present in public spaces and surveilling everyday conduct). In thinking about practice, then, NRTs draw attention to 'the background' of social life, and how:

> While we do not consciously notice it we are always involved in and caught up with whole arrays of activities and practices. Our conscious reflections, thoughts, and intentions emerge from and move with this background 'hum' of on-going activity.
>
> (Anderson and Harrison 2010a: 7)

The core of this chapter discusses a series of key terms from within this work on practice. This will start with a discussion of 'performativity' and 'the event' and, with that, a reflection on NRTs' concerns with the production of norms, becoming, and difference. Picking up on the emphasis on change and becoming in such discussions of the event, and the risk of them moving too far from a concern for the contextual nature of practices, the chapter will then turn to work on rhythm and habit which have attempted to think about the situated nature of practices. In the final main section of this chapter, I will discuss critiques of the emphasis on activity and generativity present within much of this work. This will show how bodies are not necessarily always actively enrolled in the 'enactment' of the world and that such vulnerability can be understood to constitute the condition for the possibility of such enactment.

Practice theory: from social to radical constructivism

> Practices are productive concatenations that have been constructed out of all manner of resources and which provide the basic intelligibility of the world: they are not therefore the properties of actors but of the practices themselves ... Actions presuppose practices and not vice versa.
>
> (Thrift 2008: 8)

To understand the significance of the conjunction 'Non-representational Theories and Practice', it is important to understand broader shifts taking place within the social sciences and humanities around the time that NRTs were first engaged with by geographers. The 1990s saw a 'practice turn' in the social sciences and humanities. As part of this, the work of a diverse range of practice theorists – both social theorists and philosophers – came to the attention of social scientists, including human geographers (see Schatzki 2001; Thrift 1996). Within this diverse practice theory, there are a range of subtle differences in specific understanding or target of concern. A key feature that holds this work together, though, is its contrast with, and critical disposition toward, social constructivism.

As discussed in Chapter 1, social constructivism had a significant influence on human geography during the 1980s and 1990s when it came to the focus of a range of work in New Cultural Geography. A central facet of social constructivism is a concern for representation and particularly "the structure of symbolic meaning" (Anderson and Harrison 2010a: 4). Social constructivism starts with representations and considers their role in shaping the conduct of individuals, groups, and societies. Such representations can take a range of forms from paintings to photographs to written texts to sets of texts that come together to constitute a discourse. In considering their structure, the focus is on the way that the symbolic meanings manifest in and through such representations are constructed by society. Importantly, social constructivists argued that, in being constructed, such symbolic meanings and their ordering were arbitrary, invented rather than being something that was essential and everlasting. This constituted one of the central critical insights of social constructivism (Anderson and Harrison 2010a). With the recognition of this constructed, arbitrary nature, social constructivism also recognized the possibility for there to be multiple symbolic orders. There can be social and cultural differences between what is represented, who represents, and how that comes to be interpreted, understood, interiorized, and so on between a host of people and places. And with that, it becomes possible for such meanings to change or be changed.

NRTs seek to move beyond social constructivism in their attention to practice. Social constructivism's focus on representations:

> means that 'action' is not in the bodies, habits, practices of the individual or the collective (and even less in their surroundings), but rather in the ideas and meanings cited by and projected onto those bodies, habits, practices and behaviors (and surroundings). Indeed, the decisive analytic gesture of social constructivism is to make the latter and expression of the former.
>
> (Anderson and Harrison 2010a: 5)

As such, although social constructivism "pointed us towards events and processes marked by their mundaneness and ordinariness – in short, their everydayness – its commitment to theorizing and researching social practices in their own right has been patchy" (Latham 2003b: 1901).

Practice theory approaches social life in a radically different way when compared with social constructivism. For such practice theories, representations are not the way that we know about our place in the world but rather play a part in broader socially negotiated, practical processes of knowing (Thrift 1996). Such knowing might be more about the 'intelligibility' of a situation – that it in some way makes sense – but not necessarily about a more consciously reflected upon and articulated meaning that is attributed to it. These practice theories all in some way "deny the efficacy of representational models of the world, whose main focus is 'internal', and whose basic terms or objects are symbolic representations" (Thrift 1996: 6). Instead, practice theory focuses more on the 'external', and so on the ways "in which basic terms and objects are forged in a manifold of actions and interactions" (Thrift 1996: 6).

This is not to say that representations are irrelevant or that human beings do not at times imagine, depict, deconstruct, and so on. As Thrift (1996: 10) notes, "Sometimes we frame representations. Sometimes we do not. But the practical intelligibility [our largely unthought getting on with things] is always there". Furthermore, when we do represent things, these representations are of a particular kind. As Thrift (1996: 10) continues, "the kind of representations we make are only intelligible against the background provided by this inarticulate understanding". Or as Taylor puts it:

> In the mainstream epistemological view [here social constructivism], what distinguishes the agent from the inanimate entities which can also effect their surroundings is the former's capacity for inner representation, whether these are place in the 'mind' or in the brain understood as a computer. What we have which inanimate being don't have – representations – is identified with representations and the operations we effect on them. To situate our understanding in practices is to see it as implicit in our activity, and hence as going well beyond what we manage to frame representations of. We do frame representations: we explicitly formulate what our world is like, what we aim at, what we are doing. But much of our intelligent action, sensitive as it usually is to our situation and goals, is usually carried on unformulated. It flows from an understanding which is largely inarticulate … Rather than representations being the primary focus of understanding, they are islands in the sea of our unformulated practical grasp of the world.
>
> (cited in Thrift 1996: 9–10)

In terms of more affirmative commonalities, it is possible to identify a number of recurrent themes in these practice theories which have appealed to those developing NRTs. Common in such work is an understanding of "practices as embodied, materially mediated arrays of human activity centrally organized around shared practical understanding" (Schatzki 2001: 11). This means that in many ways, the sort of practice theory that NRTs are interested in is more 'radically constructivist' in nature (Anderson and Harrison 2010a). Breaking Schatzki's summary down, we can see three key points of focus:

-   Embodiment. Practices take place through embodied, situated actions, and interactions. To engage in a practice is to engage the body in movement in and through space and time. Bodies here have the odd distinction of being both the vehicles through which practices are perceived – we sense our surroundings – and, at times, the object perceived given the body's interactions with the world around it (Thrift 1996). Practice theories often attend to a range of sensory experiences not just limited to the visual. They include the sonic, haptic, olfactory, and so on which present us with an array of 'data' about our situation in the world. Equally though, the body is often taken to be a site of disciplinary practices and so is in some ways constituted by the specific social situation it participates in (Schatzki 2001).

- *Materiality.* Such embodied practices overflow with a variety of non-human as well as human entities. These objects both constrain and afford various forms of action, acting as (at times literal) scaffolding for these practices. There is such a ubiquity of non-human materiality that it is quite often difficult to distinguish bodies and objects. For Thrift (2008: 10), "the human body is what it is because of its unparalleled ability to co-evolve with things". Often, non-human objects come to act as technological prostheses for the human body, expanding what it can do through the production of various 'hybrid' bodies. And for some practice theories, such non-human entities do not just constitute a supportive backdrop for human action but are actually active in the unfolding of practices themselves.

- *Shared understand.* Shared understanding is seen to be significant as it is through the production and reproduction of shared know-how that social life comes to persist. Practices come to "constitute our sense of the real" rather than this coming from some prior image, either mental or depicted in various texts and discourses (Thrift 1996: 7). This is quite different to the linguistic focus of a lot of social constructivist work where texts or other representations do the work of such reproduction. Rather, here, our practices make the world and that world, in turn, solicits certain forms of practice from us. Or as Anderson and Harrison (2010a: 6) put it, "meanings and values emerge from practices and events in the world".

To try to clarify some of this relatively abstract discussion, we can return again to the scene that this book started with. In the summary provided, we can see a whole host of embodied, shared, material interactions taking place in the life of that setting. For example, look again at the figures crossing the road. How likely is it that these individuals actively or consciously thought through the process and protocols that exist around using such a crossing? How many consciously went through the steps of walking to the specific crossing? How many thought about pressing the button so that their desire to cross was registered; looked at the image of a man, lit in red, and consciously interpreted that as (a) a general representation of a figure wishing to cross a road and (b) reflected that the color red meant something here – an instruction to stop; or waited attentively looking at that red symbol awaiting change, further interpretation, and then identified and carried out the action that a change would allow? Is it not more likely that such a process was learned a long time ago and reiterated through a whole host of practice, schooling, and repetition which further ingrains the specific protocols that exist here? And where this breaks down – where someone disobeys by running a red light, running across where the gap wasn't quite enough, and so on – we see varying degrees of remonstration and reprisal (rolled eyes, muttered comments, beeping of vehicle horns, and so on).

Is it not more likely that many of these individuals went through the process of crossing without thinking of much of anything at all? Perhaps, they did it entirely unthinkingly while reflecting on something else – the night before or the day ahead. Perhaps they were being affected by the music playing through their headphones which actively shaped their mood or disposition while crossing, adding a spring to their step or particular tempo to their gait. In such situations, bodies here perceived a certain sensory stimulus – in this case music – and their actions unfolded in light of that encounter. But equally, others might have felt a great effort in moving their bodies across the street in time with the rhythm of the crossing and traffic; bodies in such instances could have become perceived as objects themselves, as heavy, unwilling, a little unsteady, and so on.

Finally, throughout all of the above, we have a host of material objects and agencies that play a part in the unfolding of this scene. There is the crossing itself – a combination of paint, buttons, switches, lights, cables and poles, electricity – which plays a role in the unfolding of the practice of crossing the road. The bodies moving through this scene interact with this material infrastructure largely unthinkingly, attuning their bodies and their movements to its functioning. And, in some cases, these crossings carry on with or without the presence of human bodies; electricity flows to traffic lights that cycle through their reds and greens day and night, indifferent to the presence, or lack thereof, of crossing bodies or passing vehicles.

This all might sound quite prosaic and so of relatively limited interest to scholarship in the social sciences. However, as Anderson and Harrison (2010a: 7) note:

> If thinking is not quite what we thought it was, if much of everyday life is unreflexive and not necessarily amenable to introspection, if … the meaning of things comes less from their place in a structuring symbolic order and more from their enactment in contingent practical contexts, then quite what we mean by terms such as 'place', 'the subject', 'the social' and 'the cultural', and quite how 'space', 'power' and 'resistance' actually operate and take-place, are all in question.

In thinking through such implications, I will now turn to a series of key themes that have appeared across the literature engaging with NRTs that variously place emphasis on the practice-based nature of social life and which, at the same time, significantly de-center any introspective subject that might be assumed to be the author of such situations.

Performativity and/or the event

It is from the active, productive, and continual weaving of the multiplicity of bits and pieces that we emerge … Equally, it is from such active, productive and continual weaving that 'worlds' emerge.

(Anderson and Harrison 2010a: 8)

One conceptual couplet that has been significant for NRTs in thinking about practice is that of 'performativity' and 'the event'. These have been discussed both together and apart, but in various ways show clearly NRTs' concerns for a practice-based understanding of social life. This section will look at some of the ways that performativity and the event have been considered and how together they offer an understanding of how social life is constructed and reconstructed in practice.

The performativity of practices

'Performativity' is perhaps most commonly associated with the writings of Judith Butler and her accounts of gender and sexed identity (Butler 1993, 1999). Butler argues that gender is something that is done rather than something that we have, that we 'do' our gender rather than 'are' our gender. Such doing, though, is not something that we undertake freely. We do not get up and consciously decide our gender identity for the day as though we are taking a costume off a rack. For Butler (1999: 33):

gender is not a noun, but neither is it a set of free floating attributes... the substantive effect of gender is performatively produced and compelled by the regulatory practices of gender coherence.

Such 'compelling' takes many forms but is often pervasive. From the moment we are born, we are called into an identity; with the statement 'it's a boy' or 'it's a girl' a whole host of normative frameworks are applied or written onto the body (the color of clothing worn, the sorts of toys played with, the sorts of games or sports that are deemed proper to participate in, and so on). In such regulatory practices, Butler is both concerned with how identities come to *appear* natural or essential through the repeated citation of certain discursive norms around gender roles, and, given this constructed nature, how this might be deconstructed through subversive repetitions. As she continues: "Gender is a repeated stylization of the body, a set of repeated acts within a highly rigid regulatory frame that congeal over time to produce the appearance of substance, of a natural sort of being" (Butler 1999: 43–44). The important point here is 'appearance'. Such identities lack the depth they claim to have. It becomes possible, then, for appearances to change, for various acts to be undertaken that draw attention to gender's constructed nature (see Box 2.1).

BOX 2.1 PERFORMATIVITY AND SUBVERSION

As well as providing an account of how gender identities are produced through repeated acts within a specific regulatory framework, Butler's account of performativity is also well known for its discussion of how gender might be done differently and so how both the constructed nature of such norms might be shown and subverted. Butler highlights how through certain acts, gestures, and (often expected) desires, the *effect* of an internal 'core' of a person's identity is produced. However, this is only an inscription on the body's surface; gender is performative rather than expressive of some internal subject or depth. Therefore, Butler argues that it must be possible to 'act' a gender in ways that will highlight this artificial constructedness of gender identities. As Butler (1990: 271) suggests:

> if gender is instituted through acts which are internally discontinuous ... the possibilities of gender transformation are to be found in the arbitrary relations between such acts, in the possibility of a different sort of repeating, in the breaking or subversive repetition of that style.

Perhaps best known of Butler's discussion of such a 'different sort of repeating' is the example of 'drag'. Most commonly, 'drag' entails a member of one sex dressing as another, but in a particularly exaggerated manner. This might mean a man dressing as a woman but with certain exaggerations. This might include heavy make-up, very large hair, large fake breasts, a revealing dress, very high heels, and so on. The effect is to make it clear that this is a man and not to 'pass' as another gender. While potentially dismissed as frivolous, drag is put forward by Butler as an example of a subversive practice that can be effective in highlighting the performative nature of

---

gender identities. This comes down to the exaggerated nature of the repetition – it draws attention to the potentially ludicrous nature of a range of gender norms, expectations, and ideals. When amplified and translated into a different context, their unfounded status starts to become apparent. That said, it is important to note that "there is no necessary relation between drag and subversion, and drag may well be used in the service of both the denaturalization and reidealization of hyperbolic gender norms" (Butler 1993: 125). In this sense, drag isn't just about dressing as another's gender. It is possible for someone to 'drag-up' as their own gender. Glamour models or male strippers, for example, clearly aspire to look hypersexualized and hyper-feminized/masculine. Oiled hairless chests and ripped abs or pouting filled lips and 'photoshopped' waists, all attest to a clearly constructed, often aspirational gender norm that is meant to be desired or seen as desirable.

A limit here, though, is that "Butler is first and foremost a theorist of the symbolic register" (Thrift and Dewsbury 2000: 413). As a result, "She has very little to say about how symbolic norms are related to other social and political structures" (Thrift and Dewsbury 2000: 413). Butler's account of performativity – certainly in her early and most influential texts on this – resides within a social constructivist framing. Or as McNay (cited by Thrift and Dewsbury 2010: 413) puts it, Butler's work "lacks a description of how the performative aspects of gender identity are lived by individuals in relation to the web of social practices in which they are embedded". Further, there is also relatively little room for space in Butler's account, for a concern with the *where* of such social practices. As Thrift and Dewsbury (2000: 414) argue, "the make-up of space is crucial to who attends and what is".

As a result of such issues, a range of geographers have turned to other sources of inspiration in developing accounts of the performativity of situated practices. In particular, there has been a concern for both the ways in which practices take place in relation to various social structures that they themselves (re)produce, that practices are situated and embodied, and also that these practices are experiential – that bodies participate in them, subjectivities and sense of self and surroundings emerge from them, and a host of felt experiences circulate around this (see Simpson 2008).

A key reference point in the context of thinking about practices as performative is Paul Harrison's (2000) account of 'making sense' in everyday life. Harrison (2000) develops an account of how 'sense' – the general legibility of the world, that it is in some way comprehensible – is made through embodied practices. In line with the sorts of practice theory accounts above which are critical of social constructivism and its emphasis on representation, Harrison (2000: 497) argues "that before and within our considerations of representation, of significant things, there are processes operating through our distracted, 'tactile' knowing". Starting from representations focuses too much on the fixed, the already completed and so fails to apprehend "the lived present [understood] as an open-ended generative process; as practice" (Harrison 2000: 499). While meanings are written onto bodies, this is not solely the situation of the body. Stopping at inscription risks relegating practices to a secondary position rather than practices being understood as a generative source of change. Again here, so much of everyday life is made up by 'unthought' embodied action. And the performative function of such actions – the fact that they do something in their taking place – means that things are never quite as stable or secured as they seem or are presented to us. Instead, living in the world here means being amid constant reproduction and, often infinitesimal, change. Things happen that make sense and make subtle differences.

To illustrate this, Harrison draws on the example of losing a wallet. Harrison (2000: 503 [citing Varela]) asks us to:

> Imagine you are walking down the road with nothing much on your mind, perhaps it is the end of the day ... You put your hand in your pocket and realise that your wallet is missing.
>
> Breakdown: You stop, your mind-set is unclear, your emotional tonality shifts. Before you know it, a new world emerges: you see clearly that you left your wallet in the store... your readiness-for-action is now to go quickly back to the store.

In such an event something happens. There is an interruption to the background order of our everyday navigation of this space at this time. We move from a form of 'relaxed readiness' (walking along with a blank mind, navigating the street without thinking much about it), through a 'breakdown' (the realization that something isn't right), an emerging sense of what is different starts to become clear (the realization that you've left your wallet), and then a new order and practical orientation emerges in light of this (that you're going to go back to try to reclaim your wallet). This is an embodied reaction that eventually comes to be understood upon reflection. Something 'feels' wrong but you can't quite put your finger on it. Once you have, the potential for anxiety, adrenaline, and panic sets in. Your embodied disposition, your 'emotional tonality', shifts radically. Later, this is resolved (or not) depending on the status of the lost wallet. Such feelings come to shape our sense of self rather than belong to us.

In thinking through this processual understanding of the world where interruptions can and do happen, Harrison considers how this sense of the world is held together through practices of habit and inhabitation and that it is such habits/inhabitation that are disrupted here. Through the unfolding of everyday practice, memories come to be 'folded' into the body. This doesn't mean that meaning becomes inscribed on the body but rather that certain attitudes and competencies accrete which enable us to get on with things; a sense starts to become clear provisionally. But equally, habits can be interrupted and disrupted through sensations that open up alternative worlds. As Harrison (2000: 512) notes:

> Given the intervention of an event (a stuttered word, a loss of footing, a door slamming, a misplaced wallet, a newspaper not delivered), random lines can be thrown out from a form of life, a swerve occurs; 'something has happened', 'how did we get here?'.

Or in other words, it is the case that "in the everyday enactment of the world there is always immanent potential for new possibilities of life, which may open new spaces of action" (Harrison 2000: 498).

The unfolding of events

From these discussions of the performativity of practices, we get a sense of how social structures, meanings, sense, identities, habits, sensibilities, and so on are not pre-formed but rather come to be formed (and re-formed) through practice. This, in turn, shapes 'us' in who we are and who we (continually) come to be. Where Harrison's (2000) account leads, though, is an understanding where "Feelings, experiences, and sense are never 'owned' by a person, rather they are impersonal events and encounters" (Harrison 2000: 514). Such an event-based approach to practice has become a prominent trope within NRTs. As Latham (2003b: 1902) notes:

> Increasingly, human geographers and other social scientists are recognizing the need to acknowledge the event-ness of the world ... Thinking through this event-ness presents a vigorous challenge to many of our more deep-seated habits of thought. It demands modes of thought that acknowledge both the openness of practice and the importance of emotion, desire, intuition, and belief in the unfolding of our worlds.

Further, as Anderson and Harrison (2010a: 19) note:

> 'The event' has been such an important concept and empirical concern for non-representational theories because it opens up the

question of how to think about change … thinking 'the event' is of importance because it allows the emphasis on the contingency of orders [found in NRTs] to morph into an explicit concern with the new, and with the chances of invention and creativity.

Thinking in terms of the event further moves the focus away from agents (or subjects) who might commonly be taken to be the source of actions and outcomes. Instead, the event is taken as the starting point and any such agents or subjects an ongoing outcome of such events (Dewsbury 2000). This, in turn, expands the range of actors who play a part in the unfolding of these events. Humans are not at the center of events. There "is a shift to thinking about how worlds may be arrayed and organized with humans, but not only humans" (Anderson and Harrison 2010a: 12).

When it comes to discussion of 'the event', there are a range of ways that this has been considered amongst NRTs. This has ranged from the infinitesimal changes perpetually unfolding in the playing out of day-to-day life – events that don't really seem that eventful or might not obviously lead somewhere – to the more marked or rarified events that bring about significant and lasting changes (compare Dewsbury 2000 with Dewsbury 2007). In this, we have anything from the ongoing modification of the 'background' of day-to-day life, a 'continual differing', to something that is much more surprising, something that breaks with what went before and might lead who knows where (Anderson and Harrison 2010a).

Significant to NRTs' initial engagement with event-based ideas was the work of the philosopher Gilles Deleuze and his philosophy of difference. Deleuze's philosophy is notoriously challenging both in terms of style and content (see Box 2.2). By way of an initial introduction to Deleuze's performative take on 'the event', Dewsbury (2000: 476–477) identifies five key aspects of his work that challenges our common-sense ways of thinking about practice and emphasize this concern with difference, change, and emergence that is not dictated by a human actor:

- Connection. Deleuze's philosophy is about relations and connections that appear between a variety of different entities (human and non-human, real and ideal, animate and inanimate). As Dewsbury (2000: 476) suggests: "One asks what something does, and how in its doing, or being thus, it connects with other things, digresses boundaries instigating new ones, whilst rejecting, separating, and recomposing others". These connections between a variety of 'things' make things happen and so breed change.
- Heterogeneity. Developing from such connection, we have difference and change. Through such

relations of proliferating couplings variant domains, capacities, objectives, and outcomes are assembled, offering up, opening, and unfolding alternative, more situated, spaces that contest the discourses of our present and possible futures, vitally giving back to our studies of life the multiplicity of the present and the alterity of the future. (Dewsbury 2000: 476)

The fundamental point here is that there isn't an obvious starting point that forms the basis for such change to take place – some kind of origin – nor is there a predictable definite destination through the unfolding of an event. Rather, there is constant variation. Things don't unfold in a linear way, moving from A to B, but rather (potentially) shoot off in a range of unanticipated directions from in the midst of things.

- *Multiplicity.* From such a lack of foundation (either as starting point or end), this difference and change doesn't add up to anything like a 'totality' or an identity. Deleuze's philosophy is 'anti-foundational'. Rather, we perpetually find ourselves "in the midst of working on, making anew, amalgamating, acting and reacting with others and with things" (Dewsbury 2000: 477). In this situation, anything resembling identity, sameness, and the like is really just a state of provisional consistency that belies a host of internal differences and potential change.
- *(against) Structure.* The aforementioned focus on difference, relation, process, and so on in Deleuze's philosophy mean that the emphasis here is on the ongoing composition of the world and its meaningfulness through practice. As Dewsbury (2000: 477) suggests, "both our thought (ideas) and action (practices) assemble the relations of human and non-human and announce the discourses through which we exchange and, through description, make our experiences meaningful". Echoing much of the practice theory discussed in the previous section, our actions shape this shared situation rather than being merely a product of it.
- *Ruptures.* Amid the practical composition of the world and the ever-unfolding differentiation that takes place within this, there are also moments where something more pointed happens, though we might not realize it at that time. There is still scope for the unexpected here, for a radical form of contingency (Dewsbury 2003). Amid such ruptures, it is possible that "something happens before one construes it as happening to oneself" (Dewsbury 2000: 477).

Throughout the above, we can see that there is an emphasis on the broader situation rather than individual actors within it. The focus on relationality brings in a whole host of things beyond individual human bodies that might be engaged in various practices. And, with that, any individual human body engaged in practice present here is not at the center of the picture. The focus has shifted from specific practicing bodies toward the broader unfolding events. Such events exceed our intentions, our actions, our desires, our hopes, and so on. Events are not personal occurrences circulating around us as individuals who originate them through our actions but rather events are both indifferent to us (they'd take place with or without us) and are constitutive of us (they affect us). In this sense, "we cannot be fixed points in the world for we become with the world" (Dewsbury 2000: 480).

BOX 2.2 DELEUZE AND THE EVENT

Deleuze's conception of the event has proved a prominent point of reference for NRTs. The challenge of this concept, though, is that its focus is not on what individual actors (or even actants) do or what actually occurs (Deleuze 2004). It is not about the practices that individuals engage in or what that might achieve. Deleuze's focus is less personal than this and is not directed toward ends. The focus is on relations and what emerges. Or in Deleuze's own words:

> "An event is an astonishing, multiplying, emissive occurrence, an intense awareness or perception of something that turns into a becoming-other ... that somehow takes place in a swarm of sensations, in a nexus of 'prehensions of prehensions.'"

> (cited in Dewsbury 2000: 487)

One way we can think about this conception of the event is through an example. Stagoll (2005) suggests a tree's changing color with the change of a season. As I write this, in the yard behind my house, there is an old pear tree that I can see from my office window. It's early October, summer is over, and the leaves of the tree have changed from green to subtle shades of red and brown. They've also started to fall, and soon the yard will be filled with those leaves. Thinking in terms of Deleuze's account of the event, the event here is not what has evidently taken place – that the tree has changed color – or what will happen next (being leaf-less):

> because this is merely a passing surface effect or expression of an event's actualization, and thus of a particular confluence of bodies and other events (such as weather patterns, soil conditions, pigmentations effects and the circumstances of the original planting).

> (Stagoll 2005: 87)

For Deleuze, it isn't a case of saying that the tree became brown or became red or that the tree has lost its leaves. This would start from a position of the tree having an 'essential' state – being covered with green leaves – and that this has been changed. Instead, Deleuze asks us to think of this in terms of 'the tree browns' or 'the tree reds' or 'the leaves shed'. This adds a dynamism to the event that isn't tied to the tree's starting point (having green leaves) or end point (being leafless). As such, "The event is not a disruption of some continuous state, but rather the state is constituted by events 'underlying' it that, when actualized, mark every moment of the state as a transformation" (Stagoll 2005: 87).

Figure 2.1 The tree browns/sheds (*Author's own*).

In sum, events do not lead from x to y from one determinate state to another. Rather, the event is unfolding. The event expresses something of the forces from which it arose, the relations that took place, and is not about the start or the outcome.

Different differences?

A recurring theme amongst the discussion of performativity and the event here has been that of 'difference'. However, this difference has been understood in a very specific way. There is a shift here from the sorts of differentiation Butler discussed in the performative reproduction of gender identities to the differing that is found in Deleuzian accounts of events. Nash summarizes this in slightly different terms. The development of NRTs for her has meant that:

a return to the body and bodily practice seems to pull in two directions: one towards understanding and denaturalizing the social differentiation of bodies through practices, and one towards a more generic and celebratory notion of the embodied nature of human existence.

(Nash 2000: 655)

When it comes to the latter, there has been a concern that 'NRT' is "reproducing bodies that are not differentiated" (Colls 2012: 432). This has led to questions about the politics of such accounts and how much they do, or rather don't, attend to issues of social difference. For example, Nash (2000: 657) is concerned that we "lose the sense of the ways in which different material bodies are expected to do gender, class, race or ethnicity differently". Or as Cresswell (2012: 102) puts it, "What is missing is a group subject marked by all the old important forms of 'identity', such as class, race, and gender". For Mitchell and Elwood (2012: 6, 9), this leads to a lack of attention to "larger patterns of systematic violence" and so a "depoliticization of events". As such, it has been suggested that 'NRT' provides something of an 'asocial' or universalist account of practices.

It is important here to note that NRTs take a different approach to difference than has been common in cultural (and social) geography. NRTs often do not start from the premise of difference being between one entity and another on the grounds of, for example, form, content, significance, and so on. It is not the case here that "difference is conditioned by and derived from identity" (Doel 2010: 121). It is not the case of identifying the difference between a and b or the source and implications of such differences. Rather, the understanding of difference often adopted by those working around NRTs is concerned with a more radical difference, one which marks out each entity as singular and itself unfolding. This is not, then, a difference between specific manifestations or collectives, but rather about a difference in itself. Here, identity (and difference between identities) is something that comes *after* difference and is not the starting point from which difference is established (Doel 2010). This leads to a range of different political questions in terms of how we might be attentive to, and affirm, such difference and differentiation rather than be concerned with distinguishing identities.

While this is not necessarily to the satisfaction of many (for example, see Cresswell 2012), NRTs have

endeavoured, from [their] earliest articulations, to open Human Geography's conception of what the political means – i.e. what counts as a proper political question – by supplementing the epistemological logic of traditional forms of social/political theory.

(Rose 2010a: 341)

The question becomes whether this is a both/and or an either/or situation. Or, it asks whether there are other ways to think through that tension. In other terms, it is a question of if and how NRTs can attend to representations and what they do, to their enactive nature. For some, the non-representational nature of practice and difference becomes something of an absolute (see Dewsbury 2011a). However, for others, we start to see the implications of certain societal framings within the taking place of practice, their contextual nature. And there are a range of interesting developments taking place here that specifically try to think between such radical takes on difference from the context of a specifically feminist or anti-racist position (see discussions of Colls (2012) on sexual difference and Saldanha (2010a) on race in Chapter 4).

Continuity in becoming?

Such critical commentary appears to have shaped the ongoing evolution of NRTs. It is possible that such accounts of the event and their focus on change, interruption, becoming, and difference lead us to lose a sense of the ways in which change is constrained or how things persist in relatively stable, consistent ways. In trying to insert more in the way of "that sense of free play which would let creativity back in" (Thrift and Dewsbury 2000: 414), this lens of 'the event' does risk going too far the other way. This means that that which constrains, renders relatively stable and durable, might fall into the background of accounts of practice. There are moments where this is taken up, though. For example, Dewsbury (2000: 487) offers the following:

> Take ... the building you walk through/within – what is the speed of flux that is keeping it assembled? It seems permanent, less ephemeral than you, but it is ephemeral nonetheless: whilst you are there it is falling down, it is just happening very slowly (hopefully). In such a world, that is incessantly bifurcating and resonating amongst the different movements of its many compositions, our subjectification is always occurring.

While we have such a concern for durations and speeds in such accounts of difference and events, we do still come across the question: "where do we find our place in the world that is continually being created in the infinite movement of the event?" (Dewsbury 2010b: 154).

There are a number of ways in which scholars engaged with NRTs have responded to this. From the start, interest in ANT led NRTs to focus on the relational-constructed nature of the orders of day-to-day life (see Thrift 1996 and Chapter 4). However, such geographers have also turned to other relational frames of reference to try to think about this relationship between stability, disruption, and change. This has included interest, for example, in ideas around 'assemblages' (see Anderson and McFarlane 2011; Box 4.2) and 'ecologies' (see Simpson 2013). In this section, I want to focus on two specific conceptual-thematic framings – rhythm and habit – which have been significant to NRTs' ongoing developing accounts of practice and specifically how a situated consistency might emerge within them.

Rhythms of practice

One way that we can think about this tension between change and continuity when it comes to practice is through rhythm (see Lorimer 2007). Very generally, rhythm refers to a prominent, regularly repeated pattern of movement. This is often associated with music or sound but can relate more broadly to a range of actions that contain within them some kind of pulse or measure. A person's gait can have a rhythm. The buildup and dissipation of traffic in cities can be rhythmic based on the arrival and passing of 'rush hours'. It has been suggested here that everyday life is crisscrossed by host of rhythms that emerge from practices but that these rhythms also give some order to those practices in their recurrence.

Rhythm has been a topic of interest for geographers for some time (see Crang 2001; Edensor 2010), especially in terms of the work of humanistic geographers like Anne Buttimer (see Mels 2004). There has been a concern for how "place-bound social practices, coded gestures, metaphorical styles, technological applications, and experiences are at the same time constitutive of rhythms that operate over a variety of spatial and temporal scales" (Mels 2004: 6). However, NRTs' interests in embodied practices have led a number of geographers working in this area to develop a less subject-centered, less obviously humanistic interest in rhythm (McCormack 2002, 2005; Simpson 2008, 2012; see Box 2.3 and Chapter 6). To see this difference, we can look to a series of key themes in the study of rhythm suggested by Mels (2004) that are both useful in seeing the relevance of rhythm to thinking about practice and in highlighting contrasts between the approach of humanistic geography and NRTs (see Simpson 2008).

- *The body.* Mels (2004) suggest that the body is significant to examining rhythm. Mels (2004: 5) notes that we "experience objects, their place and our own place with our lived-living body". In light of NRTs, this could be put is less subject-centered terms. Rather that experiencing things 'with' our bodies, we experience them as bodies. Again, we are not necessarily in charge of such encounters, and we are as much acted on as acting upon. In this sense, rhythms become entangled with our bodies. There are a host of rhythms going on in our bodies (breathing, heartbeat, etc.), and our bodies encounter a range of rhythms in their situation in the world (those of other bodies, of movements, of technologies, of the working day, etc.).

-   *The social.* Mels (2004) argues that there are more to rhythms than the body of individual experience. Rhythms extend to social space. We are often in situations where others are present, and the rhythms of others come to be entangled with the rhythms of our bodies. However, this could be characterized in terms that exceed the presence of human others. NRTs have drawn attention to how 'the social' is more-than-human in composition (see Chapter 4). In extending rhythms to social space or rather a broader sociality, multiple habitual routines accrete into a social-habitual formation in such spaces. This is composed of a fusing together of the rhythms of various individuals' pre-discursive habitual movements *and* the time-space routines promoted by a host of non-human others (technologies, objects, animals, etc.).
-   *Discourse.* Mels (2004: 6) highlights how habitualized rhythms can be "connected to particular discourses, geographical imaginations and modes of representation". Such discourses and representations tend to encourage particular rhythms and routines; they can play a choreographic function (McCormack 2005). This organizes social life, influencing how actions unfold and excluding unwanted or undesirable rhythms (or more likely, irregularities). NRTs suggest that important here is less the deconstruction of the meanings and ideals present in such discourses, imaginations, and representations – though that is of significance – and more a consideration of what such discourses *do* to the unfolding of practices (Dewsbury *et al.* 2002). What do rhythms produce and what affects and effects does this have on those participating in them?

One important point not covered here is the possibility for interruption and change. In being composed of recurrent actions, rhythms can evolve or break down. If we think about rhythm in the context of an urban street, the street solicits particular rhythms in the way it brings together lots of different bodies in a social space each with their own agenda and circumstances. Through repeated occurrences, their combination becomes routine. For example, the rhythms of cars, bodies, traffic lights, lunch hours, and so on become 'normal' and synchronized through their reiteration in the street day after day. People travel to work at the same time each day, often at similar times. Traffic systems are designed (hopefully) to adapt to these changes in flow. More trains and busses run at busier times, for example. Businesses open and close at set times depending on the day (though some stay open later or around the clock). People take breaks from work at customary times – lunch hours, coffee breaks, and so on. They also work set days and not others. People go shopping on days they aren't working, often at the weekend. There are a range of space-time routines that map out the life of such spaces (Latham 2003a).

However, sometimes something can take place that disrupts this routine unfolding of everyday life in such social spaces. We can think about the situation of encountering a street performer during a lunch hour (Simpson 2008, 2012, 2014a; see Figure 2.2). Such a performance might make an intervention into the habitual rhythms of day-to-day life here, reorganizing it, for however fleeting a time. Someone might stop what they are doing for a few minutes to watch or listen amid their lunch hour or on their way to another place (a shop, a meeting, a lecture). The encounter in some way interrupts the normal rhythm of their day. It might lead to the remainder of their time (or this section of their day) being rushed through with a more frenetic rhythm. It might mean that they become late and it disrupts the timetable of others. However, this intervention can also become habitual and everyday – the same performer might perform in the same place each day and that individual might stop each day, making space for them in their routine. This can become part of a revised routine and so rhythm of their day. The music played by this street performer might, for example, start to form a sort of signature tune to this time and this space (see McCormack 2002). Equally, this once novel interruption might fall back into the wallpaper of the street. It might just be a one-off encounter that doesn't become part of a new routine.

Figure 2.2 Disrupting rhythms (*Author's own*).

Returning to the question of difference and becoming and how consistency might be accommodated within this, we can see how rhythm allows us to see both that things are constantly moving but that there is some kind of relatively stable, potentially provisional patterning and recurrence within this. Rhythm as a lens onto practices allows us to see "the practical and performative consistency of the world, whilst at the same time paying due attention to movement at varying speeds" (Simpson 2008: 810).

BOX 2.3 RHYTHMANALYSIS

'Rhythmanalysis' has formed a key point of reference for geographers interested in practice when it comes to examinations of rhythm (compare Amin and Thrift 2002; Crang 2001; Edensor 2010; Simonsen 2004; Simpson 2008, 2012). Rhythmanalysis refers to a form of analysis proposed by the sociologist and philosopher Henri Lefebvre (Lefebvre 2004). Lefebvre is probably best known in geography for his discussion of the 'production of space' (Lefebvre 1991; see Box 6.1). However, as part of his project of critiquing everyday life in contemporary capitalist economies (Lefebvre 1992, 2002, 2005), Lefebvre wrote a short book on 'rhythmanalysis' which develops a more temporal focus in terms of the production of everyday life and space. As noted by Elden (2004a: xii), "Lefebvre uses rhythm as a mode of analysis – a tool of analysis rather than just an object of it". The idea here is that rhythm can shed light on a range of details about the functioning of everyday life.

Lefebvre develops a range of concepts to be used in his rhythmanalysis of everyday life. Centrally to rhythmanalysis, Lefebvre contrasts 'linear rhythms' and 'cyclical rhythms'. Lefebvre (2004) suggests that linear rhythms are 'monotonous' and 'tiring' and cites them as having come to prominence in the modern era in 'the daily grind'. These are rhythms that are imposed upon us in light of developments in modern society (the time of factories, work-leisure, social reproduction, and so on). Cyclical rhythms, by contrast, are cast more favorably and are suggested to originate in nature (the cycles of the day, seasons, the rhythms of the body). Generally, the linear can be aligned with the social and the cyclical with the natural. With that, the former is generally presented negatively and the latter positively. Linear repetition is "exhausting and tedious, while the return of a cycle has the appearance of an event and an advent" (Lefebvre and Regulier 2004: 73). Of particular concern, then, for Lefebvre are the ways in which linear

(Continued)

rhythms have imposed themselves upon cyclical rhythms and so how day-to-day life is increasingly organized around linear rhythms. This meant that the:

> Critique of everyday life studies the persistence of rhythmic timescales within the linear time of modern industrial society. It studies the interactions between cyclic time (natural, and in a sense, irrational, and still concrete) and linear time (acquired, rational, and in a sense abstract and antinatural). It examines the defects and disquiet this as yet unknown and poorly understood interaction produces. Finally, it considers what metamorphoses are possible in the everyday as a result of this interaction.

> (Lefebvre 2002: 49)

From this conceptual couplet, Lefebvre became concerned with what he called 'polyrhythmia'. In social life, there is not just one rhythm present. Lefebvre suggests a number of concepts to facilitate the understanding of this 'polyrhythmia'. One form of polyrhythmia is 'eurhythmia'. This is when "Rhythms unite with one another in the state of health, in normal (which is to say normed!) everydayness" (Lefebvre 2004: 15). Here, rhythms sit comfortably with each other, and they are harmonious and mutually supporting. In contrast, we have 'arrhythmia' which is "when they are discordant" or in "a pathological state" (Lefebvre 2004: 16). Here, there is a disruptive or dysfunctional interrelation of rhythms.

When it comes to application, Lefebvre suggests that we become a certain sort of observer of everyday interactions. In an apt metaphor for such a rhythm-analysis, Lefebvre suggests a physical situation of the rhythmanalyst relative to what is being observed: on a balcony which will allow a view over the space under study, but is somewhere not overly distant or detached from it. From this vantage point, the rhythmanalyst will still be able to hear the music of the city, be able to "'listen' to a house, a street, a town, as an audience listens to a symphony" (Lefebvre 2004: 22), but not be lost within its din (also see Simpson 2012).

There are, though, aspects of rhythmanalysis that sit at odds with NRTs. Given NRTs focus on emergence, the construction of the social through practice, and a diverse range of agencies and felt experiences, it is arguably the case that there are more complex relations between the linear and the cyclical than the 'interference' Lefebvre admits. The tension here lies in the specific qualifications that Lefebvre gives to linear and cyclical rhythms (monotonous, tiring and vital, adventive, respectively) in *advance* of the playing out of practices. This is all fixed in advance, and the Rhythmanalytical Project becomes one of identifying where the tensions lie in light of such pre-figured qualifications. This shuts down the potentially excessive nature of these rhythms and the practices they unfold within, the chance that linear rhythms might facilitate or enable as well as constrain. Instead, attention could fall upon the emergent affective relations between and within linear and cyclical rhythms, not just their interference. This acknowledges that different rhythms do not *just* compete or interfere, but that they constitute emergent 'affective ecologies' (see McCormack 2005; Simpson 2008).

Developing habits

While habit featured within Harrison's (2000) account of practice discussed earlier, and in a less direct way through NRTs' engagements with practice theories that focus on unthought, routine actions, recently habit has been returned to by scholars working with NRTs (see Bissell 2013, 2015; Dewsbury 2011b, 2015; Dewsbury and Bissell 2015; Lapworth 2015; Lea *et al.* 2015). A central concern of such work has been to rehabilitate habit from certain accounts which present it negatively. Habits and habitual actions have been seen as bad things in being unthought, mindless, mechanistic, non-rational, and so on (see Lea *et al.* 2015). As such, there have been arguments that we should be suspicious of habits as they are often the product of the sorts of performatively enforced norms encountered earlier in the context of Butler's work. In response, recent work on habit has sought to show that habits are fundamental to the (re)constitution of daily life and offer a range of potentials, both for carrying on *and* for doing things differently. Or as Lea *et al.* (2015: 50) suggest:

> "habit has enabled geographers to understand how cultural practices situate the human body and self within such contexts, and to ask important questions about corporeal change and continuity, the role of wider structures that situate us in the world and the relationship that we have to our bodies in a diversity of contexts".

There are a number of concerns within this re-engagement with habit that resonate with the discussion so far in terms of change and continuity. As Dewsbury and Bissell (2015: 22) note, "Habit contracts the past", it archives what has gone before, but is also "extends the present" through their repetitions. However, as there is no possibility of absolute repetition, these repetitions differ and so present us with the potential for something new happening. Or to put this in terms of a fundamental geographic concept, place:

> Don't places emerge in *habit*, through the repetition of practices and performances, itineraries and routines? Each rendition is accretive, building on the last and oriented to the next. Each rendition similar to the former but with new acquisitions introduced each time, however minute or imperceptible. Habit is then a way of appreciating that a sense of place is emergent and developmental, rather than static or authentic. Through repeated inhabitation, our sense of place can change in profound ways. As our experience tells us, the strange can become familiar; the exciting can become dull; the unseen can become perceptible.
>
> (Dewsbury and Bissell 2015: 23)

In considering the relationship between individuals and their environments, habits here are not the sole preserve of human individuals. Rather, they are also implicated in the world in which such individuals find themselves. There are a range of forces and affordances there that play a part in how habits form and reform. Further, while habit gives a sense of how the world coheres day in day out, it does more than this. Here, it is useful to draw a contrast to the work of Pierre Bourdieu which informed Thrift's (1996) initial engagements with non-representational thought. In particular:

> Bourdieu's concept of habitus has been a common way of accounting for the regularity, coherence and endurance of habits, where, through repetition, grounding norms and conventions are internalised by the bodies that are socialised into them, giving rise to a practical intelligibility that shapes future practices.
>
> (Dewsbury and Bissell 2015: 24)

However, this recent engagement with habit is concerned with more than just this stability. Focusing more on what Dewsbury and Bissell (2015) refer to as the 'intensive' qualities of practice here – for example, the subtle shifts in sensations that occur throughout the unfolding of the same action each time (see Chapter 3) – also draws attention to the dynamic nature of habits and the 'plasticity' of bodies. At the same time, it also draws attention to the dynamic and plastic nature of the self that is (re)constituted by those habits. In this, I am not in charge of my habits but rather I unfold through them.

Finally, in this unfolding, we adapt to what the world presents us with, be that the shock of something new (a new workplace, home, social setting, and so on) or the return to a familiar setting (a now familiar workplace, home, social setting, and so on). Rather than being about the routine, habit here becomes a process through which we develop a sense of the world.

By way of an illustration of these relatively abstract concerns, we can look to Lea *et al.*'s (2015) discussion of 'mindfulness meditation'. In thinking through contemporary practices of wellbeing, and specifically forms of 'reflexive body techniques' that seek to work on and modify bodies, Lea *et al.* (2015: 50) show how:

> habit is seemingly founded upon the plasticity and malleability of minds and bodies, meaning that (seemingly unreflexive) habits associated with discipline, control and dominance ... contain within their very making the capacity to become undone.

In this case, mindfulness is about the formation of a subjective experience of habitual bodily practices. By looking at Mindfulness Based Stress Reduction (MBSR) and Mindfulness Based Cognitive Therapy (MBCT), Lea *et al.* (2015) explore how intervention is made into the unfolding of damaging habits (ones that might result is stress or depression). This intervention takes place through the cultivation of an awareness of such habits and potentially, but not necessarily, a different way of doing things. The starting point for these practices is that we need to become aware of our tendency to operate in our day-to-day lives on auto-pilot. This is discussed in terms of presentness – we are often not present to ourselves or 'in the present' in these moments of activity. Lea *et al.* (2015) give the example of the 'body scan', a practice whereby the participant tries to become aware of their experience in the present, specifically in terms of their embodied sensations and their commentary on this:

> the challenge is, can you feel the toes of your left foot without wiggling them. You tune into the toes, then gradually move your attention to the bottom of the foot and the heel, and feel the contact with the floor. Then you move to the ankle and slowly up the leg to the pelvis. Then you go to the toes of the right foot and move up the right leg. Very slowly you move up the torso, through the lower back and abdomen, then the upper back and chest, and the shoulders. Then you go to the fingers on both hands and move up the arms to the shoulders. Then you move through the neck and throat, the face and the back of the head, and then right on up through the top of the head.
>
> (Kabat-Zinn cited in Lea *et al.* 2015: 54)

Lea et al. (2015) draw attention to the contextual nature of such mindfulness and how this might (fail to) be integrated within or related to the specific space-time contexts of that action. It might be possible to fit meditation classes into lunch hours, for example, or even micro-breaks away from desks and the eyes of colleagues. Mindfulness exercises might be listened to during periods of travel, also. However, this also presents challenges where a literal or metaphorical space can't be carved out from the requirements of such routines, where, for example, domestic or child care responsibilities take up time, remove the chance of quiet, of attending classes, and so on. Ultimately, there is

> A complex interplay ... between the enduring habits of mind and body, the space-time routines within which they have been laid down and the ability of the participant to engage, and cultivate a different relation, with such habits.
>
> (Lea et al. 2015: 60)

Tired yet? Passivity and uneventfulness

It is clear from the preceding sections in this chapter that there is a 'busyness' to NRTs (Lorimer 2005). Rather than the world consisting of stable, everlasting foundations, everything is done and it seems to be being done all the time. It has become "the case that signification and meaning are being understood as emerging within and never departing from the contexts of practical and material action" (Harrison 2009: 989). Taken cumulatively, this emphasis on 'doings' can start to feel a bit unrelenting. Bodies just keep going, keep doing, keep enacting, keep feeling and affecting, keep iterating and re-iterating, and so on and so on and so on.... It seems that nothing is ever still. This has led some to suggest that this focus on embodied practices:

> has also allowed a series of assumptions to be smuggled in about the active, synthetic and purposive role of embodiment which need closer examination. In particular, it is assumed that bodies are bodies-in-action, able to exhibit a kind of continuous intentionality, able to be constantly enrolled into activity. Every occasion seems to be willed, cultivated or at least honed ... the experience of embodiment is not like that at all; not everything is focused intensity. Embodiment includes tripping, falling over, and a whole host of other such mistakes. It includes vulnerability, passivity, suffering, even simple hunger. It includes episodes of insomnia, weariness and exhaustion, a sense of insignificance and even sheer indifference to the world. In other words, bodies can and do become

overwhelmed. The unchosen and unforeseen exceed the ability of the body to contain or absorb. And this is not an abnormal condition: it is a part of being as flesh.

(Thrift 2008: 10)

NRTs have been criticized for always wanting to go at the ever new; at an "unalloyed 'bliss of action'" (Deleuze cited in Thrift 2008: 258). In response, Harrison (2008) sounds a note of caution. In some contrast to his earlier work which looked at the practical composition of our sense of the world, Harrison (2008: 423–424) here seeks to reflect on "corporeal existence" in terms of "its susceptibility and its passivity". This then draws attention to the susceptibility of the body, that

In their sensate materiality bodies become overwhelmed. Abilities to comprehend, invoke, and summon signification can and often do fall short, collapsing in ways which do not revert to and cannot be converted into a positive meaning and in so doing demonstrate or expose nothing more – and nothing less – than their material being.

(Harrison 2008: 425)

Bodies can, therefore, be affected by unforeseen encounters as much as be actively involved in reproducing the world. They recede and withdraw from the world as much as disclose it. Central to these points though is the argument that this susceptibility and vulnerability should not be thought of as some kind of lack or lesser situation than such (re)productive action, as something subordinate to action that needs to be corrected. Rather, passivity, susceptibility, vulnerability, and so on are "inherent in and to the living-on of all corporeal existents" (Harrison 2008: 426). This poses a range of challenges in terms of how we think about, for example, the relational composition of social circumstances, questions of agency and sense making, and the status of individual subjects in all of this.

There have recently, then, been a number of attempts to consider bodies that are passive, that are not actively engaged in disclosive action, that have suspended such active comportment toward the world around them (Harrison 2009; also see Chapter 5). For example, this has led to considerations of stillness, waiting, pain, and finitude (Bissell 2007, 2010, 2011; Romanillos 2008, 2011, 2015). By way of an illustration, we can look at Bissell's (2009b) discussion of tiredness and sleep in the context of train travel. Here, Bissell thinks about how bodies inhabit the at once fast moving but relatively sedentary spaces of the train carriage while tired or sleeping and how this (mis)aligns with the affordances offered by such semi-public spaces. In this, we hear of commuting bodies that increasingly spend time traveling between home and work given trends in employment in certain Western societies. We also hear of self-conscious tired bodies trying not to fall asleep given the feeling of vulnerability this would produce, not to mention the risk of them doing a range of socially stigmatized bodily actions like snoring, mumbling, or drooling. Such bodies also become nervous about the security of luggage and possessions as their attentive bodies shift into inattentive repose. These are bodies that both relax as they become inattentive having endured the potentially stress-filled navigation of stations, timetables, and the like, but also become anxious if their stop is part way along the line and can't be slept through. Other bodies struggle to sleep or rest in comfort, given the specific, potentially uncomfortable, materialities of the carriage (hard seats, limited leg room, and so on), noises coming from other passengers (from phone, mp3 players, or conversations), or the jerks and jolts that might take place as the train makes its way from a to b to c. And, there are a host of states between waking and sleeping which mark varying degrees of attention and inattention to the unfolding of the journey that are variously promoted or discouraged by such carriage design, socialities, and starting and stopping. This all folds into the everyday practice of commuting and are potentially integral to bodies being able to get on with life and not some kind of deficiency to be rectified through more actively reflexive activity.

Conclusion

Thought is placed in action and action is placed in the world. This is the starting point for all non-representational theories.

(Anderson and Harrison 2010a: 11)

This chapter has looked in some detail at the origin and nature of NRTs' focus on practice, both in terms of practice constituting an analytical starting point but also by hinting at some of the ways that practice has formed a focus for empirical inquiry. Having introduced practice theory's critical disposition toward the sort of social constructivist ideas that have had a significant influence on cultural geography and, with that, a focus on embodiment, shared understanding, and materiality, this chapter has look at a number of cognate terms such as performance/performativity, events, rhythm, and habit. This identified something of a shift from accounts of practice, performance, and performativity which were concerned with how the social world comes to consist and be reproduced (at times quite restrictively) in light of repeated actions toward an emphasis on a more-than human situation where a perpetual differing unfolds that human subjects might find themselves caught up in. In this, it is clear that NRTs have drawn attention to the often unthought ways in which we get about in our day-to-day lives and how this both congeals into some kind of consistent social order or normalcy but also, in being practically produced, retains within it some flexibility or opportunity for change.

That said, there have been a range of different point of emphasis across this concern for practice and critiques have emerged around them. For example, concerns have been raised over both if and how NRTs attend to the specific context in which practices take place or whether bodies float rather freely within such social situations and the sorts of different circumstances that might exist within that for differently situated bodies. Further, we can call into question the quote above from Anderson and Harrison, given the critiques of a range of work associated with NRTs over their (over)emphasis of action and activity. Not all of our embodied existence is activity, and such moments or periods of passivity and vulnerability are not problem states to be moved on from but are in fact a constitutive part of our daily existence. Recently, a number of responses have emerged to this – for example, as seen in the discussions of habit, rhythm, and passivity here. That said, a number of commentators on 'NRT' remain to be convinced, and, with that, a range of questions remain and work to be done in taking such critical avenues further.

Further reading

Thrift et al. (2010) provides an overview of the evolution of NRTs, starting with practice but then moving off in different directions toward concerns for events and life. Coming around 15 years after the writing of some of the initial introductions to NRTs, this interview provides a retrospective account and so gives a clear narrative on what was going on behind that development. **Latham (2003b)**, a short editorial introduction to a themed issue of Environment and Planning A on performance, provides an accessible introduction to the turn to practice in geography emerging from NRTs. This provides a useful context for such work, specifically focusing on the experimental approach it can lead to, the (potentially unusual) style of work practice approaches can adopt, and how the empirical world might be accounted for in light of this changed perspective. **Simpson (2008)**, while focused on ideas around rhythmanalysis, considers a range of points around the dynamism of practice and how practices take place in relation to a specific context that informs their unfolding. This is also illustrated in relation to the case study of street performance, which is shown to be both a creative/generate practice in interrupting people's everyday routines but at the same time one that happens within a specific regulated social context.

3

NON-REPRESENTATIONAL
THEORIES AND AFFECT

Introduction

affects are public feelings that begin and end in broad circulation,
but they're also the stuff that seemingly intimate lives are made of.
They give circuits and flows the forms of a life. They can be experienced as a pleasure and a shock, as an empty pause or a dragging undertow, as a sensibility that snaps into place or a profound
disorientation. They can be funny, perturbing, or traumatic ... they
can be seen as both the pressure points of events or banalities suf-
fered and the trajectories that forces might take if they go unchecked.

(Stewart 2007: 2)

Having reached this point in this book it is likely that you have responded in a
range of ways to what I have written. Hopefully, something that you have read
will have set off some kind of chain reaction whereby you have, for example,
better understood some of these ideas and, from that, can now write part of
the assignment you are working on, clarified a point in a paper you are writing, or that you have an idea for your research project. There may have been
moments of pleasure in this reading as something snaps into place. It might
just be that things previously unclear now make more sense. This might have
spurred some kind of feeling of motivation, relief, or satisfaction in moving
things forward, even if incrementally. It might be that you can now do something that you previously could not. That progress feels good. In turn, this
might mean that the pages have started to turn a little more quickly and you
are reading enthusiastically in the hope of more of these affects taking place.
But I might be being overly optimistic. Perhaps, having read the preceding
pages, you feel frustrated by this thing called 'NRT' (or is it NRTs?) and its elu-
siveness, agitated by a persisting lack of clarity as a result of failures either on
my part or in the thing itself. There may have been some 'undertow' dragging
you down, slowing you down, or disorientating you as you try to comprehend

the points I've been trying to explain. Or you might be plain pissed off by what you've read. You might see failings in my account of NRTs. Or it might be down to NRTs themselves. NRTs and/or my account of them might have produced a near visceral reaction in you. And I'm certain that some of you may be experiencing a complete lack of response, finding these pages dull and uninspiring. It might be that time, words, and paper seem to stretch out in front of you, each turn of the page bringing you yet more words to wade through. You might start flicking ahead to check how much more is left to go in the book (to save you the trouble, you are approaching 30,000 words in and there's the best part of 70,000 to go... [sorry]). You might give up – there really is probably something more enlivening on Netflix right now... – or you carry on, mechanically with coffee, Red Bull, or Monster in hand, given the looming deadline, seminar discussion, or reading group/supervision meeting you are faced with.

Such shifts in bodily dispositions and capacities that emerge when we encounter some 'thing' in the world have been a key aspect of geographers' engagements with NRTs. Work here has been concerned with how "the world is made up of billions of happy or unhappy encounters" (Thrift 1999: 302). These encounters have been discussed in terms of 'affect' and how bodies have various shifting capacities to affect and be affected by each other depending on how they come to encounter each other. This has drawn attention to a range of ways in which geographers might cultivate "a mode of perception that dwells in the midst of things" (McCormack 2007: 370). Throughout this, affect has featured in both conceptually driven writings, and it has formed a core concern of many empirically orientated studies emerging from NRTs. Affect has also come to constitute a key source of debate given the range of conceptualizations that exist here and the tensions that exist amongst them.

In navigating this sensitive terrain, this chapter will start by providing a brief overview of some of the key features of NRTs' understandings of affect. From there, the focus will turn toward a host of work developing and drawing on affect to think about a variety of geographic contexts and practices. First, this chapter will consider how these ideas around affect have acted to animate geographers' accounts of the everyday, showing the affects that circulate around, for example, listening to music (Anderson 2005, 2014), taking the bus (Wilson 2011), or just being comfortable (Bissell 2008). This section will also consider critiques of such work on affect, specifically those related to affect's argued universalism (Tolia-Kelly 2006a). Second, the chapter will turn to work concerned with the collective nature of affects and how they come to be felt and transmitted between and amongst a range of co-present bodies. Here, the idea of there being 'affective atmospheres' which engulf bodies will be introduced. Finally, the chapter will turn to concerns with how affects mediate between bodies and are mediated by other agencies. In particular, this will look at how affect can become an (elusive) 'object-target' for intervention by various authorities in various settings (Anderson 2014).

What is affect?

Affect is not an easy thing to define. As Anderson (2014: 5) suggests

> 'affect' has been used to describe a heterogeneous range of phe-nomena that are taken to be part of life: background moods such as depression, moments of intense and focused involvement such as euphoria, immediate visceral responses of shame or hate, shared atmospheres of hope or panic, eruptions of passion, lifelong dedi-cations of love, fleeting feelings of boredom, societal moods such as anxiety or fear, neurological bodily transitions such as a feeling of aliveness, waves of feeling … amongst much else.

Affect relates to but is not the same as a host of other words we might use relatively interchangeably in our everyday lives like sensation, feeling, or emotion. In an academic context, it is also challenging because affect has been understood and referred to in a range of senses across a range of ac-ademic traditions. Affect has been discussed by psychoanalysts, phenome-nologists, post-structuralists, feminists, and neo-Darwinists, amongst others (see Seigworth and Gregg 2010a, 2010b; Thrift 2004b). Putting things too simply, an affect is often understood in terms of a situation where there is "a mixture of two bodies, one body which is said to act on another, and the other receives traces of the first" (Deleuze 1978). Here, 'body' refers to a wide range of things, not just human bodies. These can be animate or inanimate, material or less concrete (i.e. an idea). Importantly, these affects are relational; they unfold in the encounters that take place between bodies (Anderson 2006). From this, bodies and their affective capacities come to be defined from these relations rather than prior to them; capacities to af-fect and be affected are relative and emergent (Seigworth and Gregg 2010b). They are also dynamic in nature given the role affects have in modifying the bodies that undergo them. Such dynamism unfolds and varies across a range of durations, ranging from the moment-by-moment to, for example, the human lifecourse (see Barron 2019).

A common reference point for NRTs when it comes to understanding affect has been the work of the philosopher Brian Massumi (see Massumi 2002). Massumi (2002) refers to affect in terms of a particular type of movement that takes place for those bodies that encounter one another. This refers to what he calls 'intensive movements' (see Keating 2019a). Rather than 'extensive' movement – a movement from one point in space and time to another (i.e. between one dot on a map to another, from my chair to the door, and so on) – Massumi understands 'intensive' movements to refer to something altogether different. Intensive movements do not relate to determined points in space or time that can be 'indexed' against such parameters. Rather, the emphasis is on the process of transition itself that bodies undergo when one 'impinges' on the other and, importantly, how such changes vary those bodies in terms of what they can do. Affects "arise in the midst of in-between-ness: in the capacities to act and be acted upon" (Seigworth and Gregg 2010b: 1).

Deleuze (1988), another key source of inspiration when it comes to work on affect, discusses such modifications in what the body can do as being conditioned by two fundamental affects: joy and sadness. Joy refers to a positive affect of an encounter, a nutrition, an expansion in our capacity to act. Sadness refers to a negative affect, a poisoning, a reduction in our capacity to act. Such affects then mark (at times very subtle) transformations:

> sometimes they weaken us in so far as they diminish our power to act and decompose our relationships (sadness), sometimes they make us stronger in so far as they increase our power and make us enter into a vast or superior individual (joy).
>
> (Deleuze and Parnet 2006: 45).

It is important here that affect is talked about in terms of *bodies* and what they can do, how these bodies are transformed, how these bodies constantly undergo such processes of variation. Echoing Chapter 2's discussion of 'the event', it is less about individuals, and rather, it is the processes of 'individuation' that are of concern here (Keating 2019a).

Affects understood as intensive movements that take place in and between bodies often take place without the presence of conscious awareness or reflection upon them. They have a degree of 'autonomy' (Massumi 2002) and so retain some kind of 'pre-reflective' dimension to their unfolding (McCormack 2003). Affects unfold before our deliberations upon them given the way that "not all of ... experience is registered by the conscious, reflective mind" meaning that "much of it never crosses the threshold of the symbolic or representational" (McCormack 2003: 494). The claim here is that there are a whole host of affective transformations going on in our bodies (and a host of other bodies) at any given moment that we barely notice or realize and certainly not as they are actually happening. As Anderson (2014: 86) suggests:

> "The wager is that there is a level of bodily life that people may not be wholly aware of, but that nevertheless has effects and can be and is being shaped or conditioned".

Sometimes, these transformations might be strong enough as to raise our awareness but they can also be quite subtle (Seigworth and Gregg 2010b). We do a host of things in carrying out our day-to-day activities without having to actively think about them or instruct our bodies to do them. There is an "immediacy" when it comes to "events like a smile, a movement, a gesture ... or the touch of a hand given over to the response of another" (McCormack 2003: 494–495). A whole host of things impinge on our bodies and make a (greater or lesser) difference to those bodies and what they can do on a moment-by-moment basis, all the time.

In explaining our lack of awareness of much of the affects we are caught up in, Massumi (2002) draws out a distinction in levels or orders of experience here between these intensive movements (or affects) that take place 'autonomously' of 'us' and some kind of qualification that is added to those experiences by us that 'indexes' them against a meaning that makes sense to a range of different people. The latter are more 'conscious' in nature in that they involve active interpretation, naming, and reference to some kind of social–cultural system (language, shared experiences, external reference points). The two are obviously related and feedback into one another, amplifying or dampening each other, but they are not reducible to each other. As Massumi (2002: 30) suggests:

> Intensity is asocial [i.e. not a part of shared frames of meaning or reference], but not presocial – it *includes* social elements but mixes them with elements belonging to other levels of functioning and combines them according to different logic.

One way that we can see this distinction between intensity and qualification is in the way that affect is differentiated from two key related terms: feeling and emotion. Anderson (2006) suggests that this can be further differentiated in terms of 'circulation', 'expression', and 'qualification'. Starting with circulation, affects occur *between* objects or entities. They are relational. It is these interactions (or affects) that are *felt* as intensive movements that circulate in and between these bodies. Moving to expression, feelings are affects experienced, affects registered in those bodies. In this way, affects find "corporeal expression in bodily feelings" and become manifest in alterations in a body's capacity to act (Anderson 2006: 736). Think of how our jaw might clench in anger as we are affected by an offensive comment overheard from another's conversation or how our cheeks might become flushed in embarrassment as we stumble while walking past a group of watching people. These feelings – anger or embarrassment – "always imply the presence of an affecting body" and "act as an instantaneous assessment of affect that are dependent upon the affected body's existing condition to be affected" (Anderson 2006: 737).

Both affect (as circulating intensity) and feeling (as the expression of affects in/by bodies) can be distinguished from emotions in terms of qualification. Returning to Massumi (2002: 28), he distinguishes emotion from affect in defining emotion as "a subjective content, the sociolinguistic fixing of the quality of an experience which is from that point onward described as personal". Emotion is "qualified intensity" (Massumi 2002: 28), it is affects (and their expression in feelings) categorized against some kind of social framework of meaning or classification that allows others, and the individual itself, to understand and communicate their embodied states. Therefore, emotion is related to an "already established field of discursively constituted categories in relation to which the felt intensity of experience is articulated" and therefore "conceives experience as always already meaningful even if this meaning is not immediately available" (McCormack 2003: 495).

This understanding of emotion as qualified intensity is argued to pose risks when it comes to being attentive to affects and their importance to the unfolding of everyday life. For Massumi and others, such a qualification restricts the movement of affective intensities which exist prior to such fixing and framing. Affects are reduced to such framings of meaning and significance (though see Box 3.1). Affect, by contrast, is "unqualified" and so is "not ownable or recognizable" (Massumi 2002: 28). Talking and thinking in terms of emotion rather than affect therefore risks missing something in experience. Such affects are understood to exceed the names that might be given to specific emotions that might make sense to others. Or as Kathleen Stewart (2007: 128) suggests, "Affects are "transpersonal or prepersonal – not about one person's feelings becoming another's but about bodies literally affecting one another and generating intensities".

BOX 3.1 DEBATING AFFECT AND EMOTION

In an observation written at a point when some momentum around affect had developed in geography, Deborah Thien (2005: 450) raised a critical objection to the direction this work was taking, suggesting that:

> This model of affect discourages an engagement with everyday emotional subjectivities, falling into a familiar pattern of distancing emotion from 'reasonable' scholarship and simultaneously implying that the emotion of the individual, that is, the realm of 'personal' feelings, is distinct from wider (public) agendas and desirably so.

(Continued)

Thien's (2005: 452) concern was that in moving away from a focus on 'the personal', affect was "employed here in masculinist, technocratic and distancing ways". She picked up, for example, on the language of 'engineering' deployed by Thrift in describing the importance of affect to urban living. For example, she quotes the following from one of Thrift's main calls for a greater attention to affect in geography:

> affect is more and more likely to be actively *engineered* with the result that it is becoming something akin to *the network of pipes and cables* that are of such importance in providing the *basic mechanics and root textures* of urban life ... a set of *constantly performing relays and junctions* that are laying down all manner of new emotional histories and geographies.

> (Thrift 2004b: 58 cited in Thien 2005: 453 [emphasis added])

Putting this concern in even stronger terms, Thien (2005: 452) argued that:

> The jettisoning of the term 'emotion' in favour of the term 'affect' seems compelled by an underlying revisiting ... of the binary trope of emotion as negatively positioned in opposition to reason, as objectionably soft and implicitly feminized

> (Also see Bondi 2005).

On that basis, Thien (2005: 450, 453) argued that "placing emotion in the context of our always intersubjective relations offers more promise for politically relevant, emphatically human, geographies" than work on affect because this moves to "get after or beyond humanity in all our diversity" and so "pushes us past the emotional landscapes of daily life".

There is a lot going on here and a number of important points to explore when it comes to both the charges levied and the relationship between affect and emotion more generally.

A fundamental point here is that "because geographers are not always talking about the same thing when they talk about affect and emotion, their conceptual, empirical, ethical and political emphasis will differ, and often profoundly so" (McCormack 2006: 330). Perhaps most prominently here, NRTs depart some way from humanistic geographies given NRTs' arguments that what it means to be human is, in a lot of ways, both to be in relation with a lot that isn't human and to be affected by that non-humanity. It is very unlikely that there is going to be a clear alignment here in the projects of work focusing on emotion and affect given the fundamental differences between them in their starting points. It is not surprising that Thien approaches affect and emotion differently than Thrift and McCormack (her two main points of critical reference).

Moving to specific differences in the vocabulary used, Thien argues that certain aspects of the language through which affect has been talked about is suggestive of a distancing from the emotional and the personal and instead promotes a focus on the reasonable and the public. In particular, Thien suggests that the language of engineering, networks, pipes, cables, conduits, and so on associates affect with a form of detached rationalism. This is an unusual charge when we look at key aspects of NRTs' accounts of affect and what they bring into view. In most accounts of affect informed by NRTs, affect is precisely about the *pre-rational*. This is a key point of the account of affect as 'autonomous' and 'intensive' outlined earlier. Affects unfold before any *rationalization* (i.e. qualification, to use the language of Massumi) by a subject who might come to feel those affects. Their being public, then, is a very complicated issue. Affects do not belong to a subject as they emerge from relations between bodies. In that, they are *social* rather than *public* (Kraftl and Adey 2008). Being relational does not make them *public*. Rather, they are pre-personal or trans-personal. One of the immense problems posed by trying to study affective experience is finding means through which to convey something of this affective experience given that it occurs outside of the realm of subjective, reflective experience (see Chapter 7). So, the implications that Thien reads from certain vocabularies used by NRTs when talking about affect are in some distinction to a range of things that NRTs actual say in their accounts of various affects.

Being more critical, Thien's appeal to something 'emphatically human' risks the same charge that she levies at 'NRT'. While Thien argues that emotion is not rational, the humanistic (and so universalist logic) underlying an appeal to emotions being central to our relations risks suggesting (falsely) that such emotional experiences mean the same thing to different people. We risk ending up with some sort of emotional 'fundamentalism' (McCormack 2006) or the naturalization of emotions (Anderson and Harrison 2006). This, it fact, would be more suggestive of a rationalization of experience – a leveling out between people – than work taking affect as its focus which very much emphasizes the singularity of affective experiences, that they take place differently amongst and between different bodies in light of a host of circumstantial difference.

(Continued)

Even if the characterization of work on affect as being masculinist/ rational/public can be questioned, it is fair to say that the work that Thien cites does risk presenting an either/or account of affect and emotion, with a clear emphasis on affect over emotion (see McCormack 2006). The conceptual framework outlined above around affect-feeling-emotion does risk presenting these related concepts in a hierarchical arrangement, prioritizing affect and relegating (in relative terms at least) feelings and then emotions (also see Keating 2019a). That arrangement risks diminishing the role or significance of emotions and, particularly, how they might feed back into the unfolding of affective relations in the way the predispose bodies toward certain responses. In response, Anderson (2006: 737) has suggested that we can see these emotions as "making space-time" and so as "qualifications that fold into a set of more extensive relations". This means that they can "be described as artful types of corporeal intelligence-in-action enacted from within a subtle choreography" in the unfolding of day-to-day life (Anderson 2006: 737). Rather than presenting a linear relation – one leading to the other and to the other – things are far messier than this in that "the three modalities slide into and out of one another to disrupt their neat analytic distinction", they feedback as well as feeding forward (Anderson 2006: 737). Feelings and emotions and affects are performative in terms of doing something in shaping the unfolding of the everyday.

Intensive movements in everyday scenes

Returning to the scene with which this book opened, we were presented with a range of different bodies enrolled in a host of relations, both present in that scene's spatial–temporal extent, but also beyond this – absent presences affecting these bodies and their capacities to act.

Let's focus on the bodies that we suggested might feel sad, depressed, or anxious for a moment. Such feelings could be seen to be expressions of a range of affective relations unfolding as we look at this scene. These anxious feelings might be expressions of an affective relation between the exam or test that body is moving toward, with its heart rate accelerating as that space-time of assessment comes closer. This affective response and its expression might be amplified by the increasing amounts of pressure being put on student bodies here in the UK through a combination of high tuition costs and challenging economic circumstances/job prospects that lie in the future. We could even start to wonder to what extent anxiety might present the affective tonality of our post-financial crash, austerity-politics, era. But, more prosaically, feelings of depression or sadness might be an expression of the affects generated by a text message from a partner cancelling plans or even a relationship, something felt all the more pointedly due to the mechanism of communication. Or equally, the source of such a depression might be hard to pin down – its affective (lack of) intensity might have crept and taken hold over time as a result of a whole host of relations and encounters. All of this might be shared with someone met on or after this walk (a friend, a classmate, a tutor, a doctor, a councilor, and so on), conveyed in terms of the emotions that this person is experiencing, something which in turn feeds back into their affective/felt states, either cathartic or leading to a spiral of upset/panic/malaise.

We can elaborate this further in looking to those bodies we thought could be hungover. Here, a whole host of affects are unfolding though neither we nor those bodies might be aware of their specific unfolding. Nonetheless, these relations are characterized by intensive movements and transformations in capacities to affect and be affect (Latham and McCormack 2004; McCormack 2007). Here, "the affects of alcohol are implicated in particular forms of sociality, of ways of being and relating through the urban, ways of moving, gesturing, walking and talking" (Latham and McCormack 2004: 717). Such affects can be "understood in terms of a field of distributed and barely tangible forces from which sensible economies of experience emerge" that are "implicated in the enhancement or diminishment of very mundane capacities to act" (McCormack 2007: 372). For example, dehydration disposes the body differently – bright lights and loud noises affect the body in ways that might be felt all too intensely – while a skipped breakfast from sleeping in might lead to different capacities to affect and be affected as a result of low-blood sugar levels. A heavily caffeinated energy drink or a strong coffee might, for a time at least, counter such diminished capacities or, if taken in unmoderated quantities, produce a body that is jittery and on edge. Ultimately, such affects inflect the sensibilities of bodies (McCormack 2007). In contrast to the ways in which alcohol might amplify behaviors when consumed (Latham and McCormack 2004), there might in turn either be a dulling of affective capacities and response as 'the morning after the night before' unfolds and as the fatigued body struggles through the day, or a heightened 'edginess' that comes to be expressed either through the action of caffeine on the body and/or as a result of a suffering body's encounters with people and things that rub up against it the wrong way.

Animating the everyday

NRTs' interests in affect have the potential, in a range of senses, to animate the sorts of scenes that a variety of geographers are interested in. We can start to see how "life takes place with affects in its midst; or, more radically speaking, how life is composed in the midst of affects" (Lorimer 2008: 552). This affective turn in human geography has expanded significantly the range of themes, questions, and materials that are now included in geographic accounts. This shows the world to be far more complex and lively than might have previously been recognized. In this section, I will look at a range of examples of such animations that span a host of everyday and, at first glance at least, banal contexts of activity, sociality, movement, and inhabitation.

Banal affects

It hums with the background noise of obstinacies and promises, ruts and disorientations, intensities and resting points. It stretches across real and imaginary social fields and sediments, like some kind of everything. This is why there is nothing dead or inconsequential in even the flightiest of lifestyles or the starkest of circumstances.

(Stewart 2010: 340)

Affects are happening all the time. They are everything and so everything is animated by intensive movements emerging from the relations that take place in and between bodies. The argument of NRTs here, then, is that people are "creatures of affect" and so that "The animation provided by affect is crucial in the practice of world making" (Amin and Thrift 2013: 14–15). However, while affects may well be an ever-present feature of our existence that exert a significant influence over the unfolding of our day-to-day lives, that is not to say these affects have to be grand in scale, intensity, or impact. In happening everywhere all the time, they happen in both spectacular and subtle ways.

Sound affects

One characteristic of a range of the intensive movements considered by geographers is that they are often quite banal in terms of the scenes of their unfolding or the relations they entail. Going to a club (Malbon 1999) or a live music performance (Morton 2005; Revill 2004; Wood *et al.* 2007) might immediately strike us as situations where bodies come to be bound up in a range of affective relations (see Chapter 6). Sharing an experience here as part of a crowd that might be singing or dancing along to the music played can create a particularly affecting situation. We can think of the quite visceral ways in which extremely loud sound might come to be felt in our bodies – the bass frequencies of a bass guitar or kick-drum rumbling in our chest and stomach or the treble frequencies encountered in a guitar solo cutting through us in a way that might verge on painful. Here, sound becomes physically felt as well as heard (Simpson 2009). And all of these relations might be modulated by the range of substances that might be consumed as part of such a night out (see Jayne *et al.* 2008; Latham and McCormack 2004).

However, music can also affect us in other more routine contexts. For example, Anderson (2004, 2005, 2006, 2014) has explored the ways in which affects are felt in the context of domestic practices of listening to music. Anderson (2006: 741) illustrates the animating/diminishing function of affect in exploring "how the materialities of music are bound to the routine and rhythms of everyday life through changes in capacities to affect and be affected". As part of this, Anderson (2004) shows how music might intervene (or not) into feelings of boredom and the dulling of affective intensities and capacities to affect that this often brings. We start to see how such music might enable "bodies to go on" within potentially depressing or hopeless circumstances (Anderson 2006: 744).

Taking one set of examples from Anderson's research, we can see how certain capacities or dispositions emerge from encounters with music in the home. In this, Anderson (2014: 93–94) focuses on the affects that come to circulate around feelings and dispositions of 'hope' and how "hopefulness as a constellation of specific bodily background feelings emergent from the expression of affect" can emerge from certain practices of listening and the music listened to in that. Such dispositions emerge through processes of judgment that are not necessarily thought through rationally. Rather, these judgments are arrived at through felt experiences and specific relationships around what might be listened to and the (un)suitability of that music relative to specific everyday situations and individuals' embodied dispositions within them (Anderson 2005).

Anderson (2006, 2014) gives the example of Steve – a 29-year-old recently unemployed man who lives alone in an area suffering from post-industrial decline – and how listening to certain music allows him to 'get on with it' in some way amid a landscape offering little in the way of hope or prospect. It is not so much a positivity inherent in the music listened to that brings about such affects and feelings. Rather, Steve finds solace in listening to relatively melancholy music (Smashing Pumpkins, Radiohead, etc.) that makes him feel like he is not alone; he is affected by an encounter with someone who might feel like him (or feel worse). Or as Anderson (2014: 96) surmises "Momentarily, music smoothes over despair and induces the affective presence of something better, in this case by presenting a shared experience … a minor variation in a 'force of existing'". At the same time, though, there can be a disconnect between the music listened to and the intensive movements unfolding within and through the body which acts to amplify or entrain negative affects and feelings. Anderson (2014) shows this in relation to another research participant (Emma) who recounted how listening to particularly cheerful music – demanded by her son – at a time when she was depressed following the news of a family death failed to affect her. Rather than being cheered up by the music, positively affected by its upbeat rhythms or messages, she disengaged from it, failed to feel anything. This, in turn, amplified the negative disposition she had found herself adopting. This music induced and escalated her disconnection with what was going on around her in "enacting a set of distinct feelings" (Anderson 2014: 98). Here, "Everyday life is held together, and fragmented, through the circulation and distribution of affects that enact multiple topologies of space-time" (Anderson 2005: 648–649).

Architectures of affect

Important to thinking about such animating capacities of affects, then, is the context in which affects take place. Kraftl and Adey (2008) show this in their discussion of affect, architecture, and the geographies of being in buildings. Kraftl and Adey (2008) show how architects and the buildings they design specifically target or limit certain affective experiences through the very design of the spaces those buildings contain. While such buildings might be interpreted in a range of ways given the host of symbolic meanings they can contain, Kraftl and Adey (2008: 214) suggest that there are also "many nitty-gritty, material-performative details that are so important to both architects' designs and users' experiences that evade perhaps all (visual) symbolism". On that basis, they suggest a need to focus on "the role of affect in the design and construction of effective spaces for inhabitation" (Kraftl and Adey 2008: 214). In particular, such buildings are argued to be involved in both the creation and limitation of a range of affects. This might involve fairly generic affects and feelings – homeliness, comfort, peacefulness, and so on – but also more specific affects as part of this.

In showing this, Kraftl and Adey (2008) discuss two contrasting architectural settings – a Steiner school and a multi-faith prayer room in an airport in North-West England. Kraftl and Adey (2008) show how certain design features of the school building itself – the use of natural materials, rounded corners, certain decorations, the presence of 'cubby holes', and a child-scale kitchen – are intended to produce a welcoming 'womb-like' interior for the children attending the school. That said, these designs do not just work in terms of generating certain affects by themselves. Rather, there are a host of practices which are engaged in by those present in the attempt to enact and re-enact such affects over time. For example, 'homeliness' is enacted in the child-scale kitchen through practices of 'playing kitchen' and doing 'authentically' homely activities like baking. For the airport prayer room, the specific affects intended are in some contrast to those often associated with spaces of travel like airports (also see Adey *et al.* 2013). Airports might often be seen as spaces of stress, tiredness, anxiety, or (increasingly) suspicion (see Adey 2009). However, the prayer room was specifically designed with the intention of attending to "the stress of travel and to engender a sense of peace at the airport through the production of very definite affects" – 'meditative states' and 'calm' (Kraftl and Adey 2008: 222). While the school building mentioned above sought to 'fit' its landscape, here, the prayer room was deliberately designed to sit in contrast to the terminal spaces of the airport; different affective capacities emerge through spatial distance and the definition of clear boundaries. Further, the specific furnishings of the space were chosen to evoke feelings of calm – soft colors, smooth lines, images of 'naturalistic scenes', and so on. Equally, other senses are also stimulated in an attempt to generate these positive affects: the use of potpourri and specific scented plants are included in the space, the lighting is maintained at relatively low levels, and the room does not have windows to the outside or terminal itself so as to create a feeling of being cocooned against such a potentially stressful setting.

Between these examples, we can see how:

> the capacity of a building to allow inhabitation to take place – and to create meaningful effects – constantly emerges through ongoing, dynamic encounters between buildings; their constituent elements; and spaces, inhabitants, visitors, design, ergonomics, workers, planners, cleaners, technicians, materials, performances, events, emotions, affects, and more.
>
> (Kraftl and Adey 2008: 215).

Difference in affects

As well as the specific architectural context or social setting for encounters, the societal context and forms of social relation this comes to promote can, in various ways, come to matter in terms of affective relations. There have been concerns over the 'universalism' of work on affect and arguments that such work does not pay attention to the ways in which collectives are differently capable of affecting and being affected due to various 'power geometries' that different bodies find themselves positioned relatively within (see Tolia-Kelly 2006a). However, a range of work drawing on ideas around affect emerging as part of the ongoing development of NRTs has shown this in a number of ways. For example, Dan Swanton's work has sought to show how 'encounters with difference' in the context of everyday life in a multicultural society, and particular encounters with different racialized bodies, unfold in light of a range of affective intensities through which such bodies are sorted and judged

(Swanton 2010; also see Lim 2010; Saldanha 2010a; Wilson 2013, 2014, 2017). Rather than focusing on racial difference in terms of biology or how society constructs these differences, Swanton (2010: 2335) is concerned with how race comes to matter in everyday life, particularly through a range of "loose and mobile racial summaries" that come to "stick to and arrange bodies, things, and spaces and produce the basis for rapid judgments that then form orientations, shape interactions, and direct action" (also see Chapter 6). Here, a range of not-necessarily conscious responses emerge through encounters with different bodies and the objects that might accompany them.

Swanton illustrates this through the case of false assumptions being made about four male bodies encountered 'out of place' following the failed London bombing attempt in July 2005. Here:

> Under the white gaze of Haworth's residents the affective energies sticking to some bodies were intensified. On this Monday morning there were, of course, no terrorists in Haworth – a chocolate-box village on the edge of Keighley that is probably better known as the erstwhile home of the Bronte sisters. Rather, four men – 'three of Asian appearance and one white' according to that week's edition of the Keighley News (23 July 2005, page 1) – were renting a cottage on Main Street. The passer-by pinned to the floor was an Israeli tourist, who happened to fit a description – 'he had dark skin and looked Asian' (ibid). In the end, according at least to the local press, the police apologised for the misunderstanding and even laughed about the incident with the men over a cup of tea.
>
> (Swanton 2010: 2333)

A social situation where a heightened alertness to terrorism – and so the sorts of (stereotyped) bodies that might engage in such activities – inflected the ways these four bodies affected those that encountered them. The capacities of some bodies (residents) to be affected and for the bodies of others (and the things that they were seen with) to affect (the four men, their hire car, the objects they carried with them) were transformed. Post-terrorist practices of policing, degrees of alert or threat, and mass-media reporting on these issues – alongside a host of background encounters and educations that shape the dispositions of bodies toward others – act to "educate affects, habits and dispositions" (Swanton 2010: 2340). As Swanton (2010: 2335) explains:

> The becoming-terrorist of four men exposes the operation of a racism of assemblages as loose racial summaries distributed across bodies, things, and spaces became the basis for perception, judgment, and action. Suicide bombing, Islamism state of emergency, cultural

difference, al Qaeda, segregation, and the 'war on terror', stick promiscuously to elements that included salwar kemeez, a box, drawn curtains, rucksacks, designer stubble, accent, brown skin, cobbled streets, a hire car, the Qur'an, etc. in processes of differentiation. The habitual clustering of bodies along the cobbled streets of Haworth was disturbed as these men failed to 'fit in' as either locals or tourists. Suspicion was stirred. Curtains twitched. Panic. 999. Misinformation spiraled into a terror alert and police raid.

Given such concerns with the imbrication of power and affect, and so with "how race and ethnicity emerge in social practice and interaction" in "the incipient organization of affects", we can see efforts to develop "an antiracist politics that works to create different conditions for affect, thought, and action" (Lim 2010: 2393–2394).

Mobile affects

While affects emerge in the interactions of bodies and take place as intensive movements – shifts in capacities to affect and be affected – the extensive movements of such bodies still matter. A range of geographers interested in affect have considered the affects that emerge in and through space-times of shared mobility. There have been a range of considerations of how time spent on the move with others entails a host of affective relations between different human and nonhuman bodies. Something as hum-drum as a daily commute has been shown to be permeated by a range of affective relations and felt dispositions. Various forms of mobility – taking the bus, catching a train, or travelling by airplane – bring different bodies together in space-times of encounter that unfold across a variety of durations in a range of ways (see Adey 2017; Bissell 2010).

This is something that Helen Wilson's work on bus travel shows clearly. As Wilson (2011: 634–635) notes, the bus is a site of "intense encounter" given the ways that its interior spaces are "spaces of exchange and closeness – or 'propinquity'" (also see Wilson 2017). Here, "where bodies are pressed up against each other, seats are shared, and personal boundaries are constantly negotiated", Wilson (2011: 635) argues that "we find an important and often overlooked site of ordinary multiculture, where differences are negotiated on the smallest of scales". Such relations play out in terms of affects and feelings which shape the dispositions of bodies toward each other. There are certain features of such bus travel, in the United Kingdom at least, that mark it out as quite specific in form and as important in this sense. There is not, for example, any reservation of seating meaning that this is negotiated on a journey-by-journey basis by each passenger; passengers stand as well as sit and there are certain (loosely enforced) codes on priority in accessing various seating; the bus starts and stops regularly meaning the composition of passengers changes very frequently; and, there are certain demographic trends in who use buses as a means of travel and assumptions around class and prosperity around this. Bus travel, then, presents us with "a shared experience that is continuously developed and redeveloped through the unpredictable coming together of multiple objects, signs, and people" (Wilson 2011: 637).

Such shared experiences might unfold in a range of ways. It might be tied up with various irritants we encounter here and which affect us in such a way that we come to feel annoyed or unsettled:

> Ring tones, music from headphones, crying babies, and offensive language overlap and compete with each other, fade into the background, or disrupt private thoughts to heighten nervous energies, which might be made apparent through corporeal display – an individual's fidgeting, head-shaking, hand-wringing, foot tapping, or exaggerated sighs.
>
> (Wilson 2011: 644)

Taking a more specific illustration of this affectivity in bus travel, Wilson (2011: 638) provides the following extract from her research diary, recounting a scene from within a bus where affects came to take on a quite pointed expression:

> Arms reach out over others; hands hold onto overhead rails and other bodies to prevent falling as the bus rounds a corner. Foreign languages. Laughter as a girl loses her balance and falls into the man behind her. Embarrassment. Apologies. Shared smiles. A man at the front gets off. The bus pulls away and people reshuffle. Redistribute the space. Somebody at the back wanted to get off. Anger. People get in the way. Outrage! Somebody is shouting. Tuts and exasperation as people get shoved aside. They push back. Snarls. The doors open and the man leaves. Silence settles amongst the passengers.

Here, we can see a host of intensive movements unfolding in and between bodies which come to be felt in a range of ways. Bodies contact each other and, in so doing, produce feelings of embarrassment. Bodies obstruct each other and, in so doing, produce feeling of anger. Anger subsides into a subdued silence. The motion of the bus, the density of bodies present here, amongst other things, impact on a range of bodies' abilities to act and affect/ be affected; to stand, to move, to disembark, and to interact.

Importantly, Wilson (2011) also shows that such interactions, often, will be amplified or mediated by the specific bodies present and a host of unfolding subjectivities that take shape during these journeys themselves. Their dispositions (sober, drunk, tired, 'up for it', and so on) can play a part here. Equally, their identities and the assumptions made about those identities by other passengers matter. At the end of a long day, the irritants mentioned above might just feel that bit more annoying and our capacities to tolerate them that little bit diminished. Equally, though, the affect of a person behaving rudely to another – perhaps not engaging in conversation with another passenger – might be felt differently based on the assumptions that one body arrives at the encounter with about that other. What Sarah Ahmed (2010) describes as our 'angle of arrival' into such encounters matters. In this way, there are "countless acts of misrecognition" unfolding at any given time as bodies respond to others without full reflective clarity (Wilson 2011: 643). Finding ourselves in "the throwntogetherness of bodies, mass, and matter can produce unpredictable affects" (Wilson 2011: 645). Often, we're too much in the midst of things to really see what is happening and so work as much on affects conditioned by past experience as by what is really happening. Or as Wilson (2011: 646) surmises, these spaces are "continuously (re)negotiated and (re)ordered through mobile and constantly shifting practices of movement, behaviour, expectation, assumption, and assertion as people come and go".

Not all journeys entail such overtly charged/amplified affects, though. For example, David Bissell has shown how train journeys are characterized by a range of subtle affects and felt experiences (Bissell 2008; also see Bissell 2009b). A range of affective sensibilities can be seen in train travel. As Bissell (2008: 1699) explains, "Whilst bodies may be physically still, the body may not cease to be moved, affectually". Bissell (2008) focuses on the complex configurations of comfort and discomfort that might emerge from spending time seated in such spaces. While we might think of comfort as being a property of an object – one chair being more comfortable than another, for example – Bissell suggests that comfort can also be understood as a specific form of 'affective resonance' that circulates between various bodies and objects. In this sense, "comfort is not wholly predetermined and is reliant on bodies themselves to facilitate this circulation" (Bissell 2008: 1701).

To illustrate, Bissell (2008: 1702) reflects on the differences in chair provision in train travel, suggesting that "The chair is routinely at the nexus of how feelings and affects of comfort are weaved through, comprehended, and evaluated". In first class, chairs will often be wider, softer, have more leg room, and made from materials that are more luxuriant and/or more pleasant to touch. By contrast, chairs in standard class are often narrower, harder, and made from hard-wearing plastic materials and coverings. Further, when it comes to seating provision in stations and waiting rooms, these are often made of hard-wearing materials (again plastics but also metal) that are cold to the touch, lack support, and require an upright posture. Between these situations, "The various surfaces and material dimensions ... set up a different relationality between the body and the chair" with each "designed to induce a particular set of affective relations" (Bissell 2008: 1705). In first class, relaxation is targeted. In station spaces, such designs seek to engineer a 'body-in-waiting' that is alert and attentive. In standard, bodies are somewhere in-between, potentially shifting in affective states as the journey unfolds.

Collective affects

A recurring point amongst these discussions of affect is that affects arise in and through relations. In various situations, affects emerge from 'sociality' – in relations between people, between people and things, and between things. However, this also poses questions in terms of how affects might either be felt collectively by multiple bodies present in a situation or, at the same time, how affects might be communicated between bodies. How might affects be 'transmitted' between bodies? In what ways can the "affects of one person, and the enhancing or depressing energies these affects entail, ... enter into another" (Brennan 2004: 3)? In response to such questions, geographers have turned to ideas around 'atmospheres' to think through such collective affects (Anderson 2009b; Bissell 2010; McCormack 2008a; also see Edensor 2012). Sumartojo and Pink (2019: 6) describe such 'atmospheres' as:

> a quality of a specific configuration of sensation, temporality, movement, memory, our material and immaterial surroundings and other people, with qualities that affect how places and events feel and what they mean to people who participate in them.

In this sense, while "Seemingly ephemeral, seemingly vague and diffuse, atmospheres nevertheless have effects and are effects" (Anderson and Ash 2015: 35).

Geographers here have been interested in a range of atmospheres emerging within a range of contexts and from a range of practices. This has included anything from the atmospheres emerging from states of emergency (McCormack 2015a), to moments of humour (Brigstocke 2012; Emmerson 2017), to practices of security (Adey 2014; Adey et al. 2013), to moments of nationalist celebration (Closs-Stephens 2016), to those that exist between various technological objects (Ash 2013, 2016), to those found at night-time in the city (Edensor 2012; Shaw 2014) or in the dark and light of the battlefield (Thornton 2015) or in the context of rural living (Maclaren 2018), and those that exist around circumstances of austerity and precarity (Hitchen 2016, 2019; Raynor 2017a, 2017b), amongst a host of others. Here, atmospheres often "form part of the ubiquitous backdrop of everyday life" but a backdrop that can be "forceful and affect the ways in which we inhabit ... spaces" (Bissell 2010: 272).

Atmospheres exist in an indistinct situation. Atmospheres emanate both from objects and bodies and, in that, emerge both before and alongside the formation of subjectivities that might register them (McCormack 2008a). Atmospheres are not simply objective given the differing ways that individual bodies might arrive at, register, and receive them. In turn, such bodies contribute to those atmospheres' character and unfolding through their involvements with them (Anderson 2009b; Bille *et al.* 2015). Thinking about affect in terms of atmospheres means that we end up with a particular sense of relationality. The 'voluminous' nature of space and our common situation of being engulfed by this means that our interrelation with that becomes apparent. In that:

> the atmospheric suggests a relationship not only with the body and its immediate space but with a permeable body integrated within, and subject to, a global system: one that combines the air we breathe, the weather we feel, the pulses and waves of the electromagnetic spectrum that subtends and enables technologies, old and new, and circulates … in the excitable tissues of the heart.
>
> (Dyson 2009: 17)

Thinking about affect in terms of atmospheres, then, emphasizes how our surroundings can in various ways "envelop and press upon life" (Anderson 2009b: 78) and as such show how "There is no secure distinction between the 'individual' and the 'environment'" (Brennan 2004: 6).

Such understandings mean that atmospheres present a complicated and ill-defined spatiality. As McCormack (2008a: 413) notes, atmospheres can be understood as spaces "simultaneously processual, distributed, and sensed". That is, these atmospheres present us with space-times that can either evolve moment-by-moment or come to hold some consistency; they spread through such space-times taking in a range of human and non-human entities without clear boundaries, and they come to affect and be experienced in and through various bodies. Expanding upon this, McCormack (2008a) suggests that affective atmospheres can be defined along two interrelated lines. We can view atmospheres in their meteorological sense as "a turbulent zone of gaseous matter surrounding the earth and through the lower reaches of which human and non-human life moves" (McCormack 2008a: 413). We can also think of their more specifically affective sense as "something distributed yet palpable, a quality of environmental immersion that registers in and through sensing bodies while also remaining diffuse, in the air, ethereal" (McCormack 2008a: 413). While these senses overlap and interrelate (see Ingold 2015) – the meteorological can be affective, for example – it is worth considering each in turn here.

Affective atmospheres

The affective sense of atmosphere is often understood in terms of how a space can have 'a feel' or mood. Here, atmospheres "seem to fill the space with a certain tone of feeling like a haze'" (Bohme cited in Anderson 2009b: 80). We can start to understand the appeal of talking about this in terms of atmospheres here if we look at the roots of the word. Here, "atmos" indicates "a tendency for qualities of feeling to fill spaces like a gas", and "sphere" indicates "a particular form of spatial organization based on the circle" (Anderson 2009b: 80). Atmospheres "'radiate' from an individual to another" and "envelope or surround" those present (Anderson 2009b: 80). This is often explained in terms of a person walking into a room and 'feeling the atmosphere' (Brennan 2004). At some point, we will have likely entered a room and, without quite knowing why, got the feeling that something was going on here. Perhaps, we had interrupted an argument or people talking about us behind our back. Subtle ques affect us – the tightness of a jaw, a shared look between others, the tone in someone's voice as they talk to us or another, a lowering of volume or turning of backs as a conversation is quickly concluded. Or equally, it could be more positive. The sounds of laughter, the animated nature of conversations, or the vigor of gestures can give a space a feeling of vitality for a time that might be infectious and so we find ourselves caught up in such excitement, even if we don't quite know what it is we are actually excited about (see Emmerson 2017). However, it is possible that a given space might be less coherent than this implies. We might question whether it is the case here that there is a single atmosphere register differently or varying atmospheres rubbing up against one another (see Anderson and Ash 2015).

Taking an example from my own research, I have explored the ways in which cycle commuters come to be enrolled in a range of affective relations with their surroundings and other human and non-human bodies (also see Jones 2005; Spinney 2009). Such affects produce specific felt atmospheres and so a certain perception of their surroundings (Simpson 2017b, 2019). Cycling involves a particular type of exposure when compared with the sort of mobility discussed above. While cyclists might wear certain forms of clothing and/or a helmet, they are relatively exposed when compared with various forms of motorized travel. They are not enveloped in a bubble of glass and steel (i.e. like a bus, a car, or a train carriage). This becomes significant when we consider how certain forms of transport infrastructure tend to "bring together differently mobile bodies, bodies moving at differing degrees of speed and slowness, moving along differing trajectories" (Simpson 2017b: 432). Such compositions of mobile bodies tend to produce different sorts of affective relations and atmospheres for these cyclists. Often, this is not problematic as bodies move with some synchronization. However, on occasions where differently mobile bodies – cars, buses, motorbikes, cars, lorries, and so on – encounter each other in some proximity, the atmospheres generated and affects felt by cyclists can trouble (see Figure 3.1). Such encounters can be staged by certain forms of infrastructure given the way that they bring these bodies together. Narrow streets, shared bus-cycle lanes, shared pedestrian-cycle lanes, traffic calming islands, and so on can bring differently mobile bodies together in a variety of ways that can become less than convivial or harmonious in tone (see Figure 3.2).

The car in the image has just passed very close to the cyclist when overtaking. The affect this proximity produced is manifest in the cyclist's gesture towards the motorist.

Figure 3.1 Proximity affects (*Author's own – from Simpson 2017b*).

In the image on the left the cyclist approaches a shared pedestrian / cycle path. The pedestrian ahead has just disembarked from the bus at the side of the road. As the cyclist approaches the pedestrian (image on the right), the pedestrian looks up and makes an inaudible comment. In the interview the cyclist commented that it sounded like the comment included a reference to 'cycling on the pavement'.

Figure 3.2 Dirty looks (*Author's own – from Simpson 2017b*).

Meterological atmospheres

Thinking in the meteorological sense, we can start to see how there are a range of materials that affect us but which might not be immediately tangible or visible to us (see Anderson and Wylie 2009; Chapter 4). Ingold (2007) talks of this in terms of the 'weather-world' so as to draw attention to the ways in which we do not just relate to other objects or entities on the earth's surface. This leads to a shift in our understanding of our position in the world. Rather than living on and moving across surfaces as 'exhabitants' of the world, we are immersed in a turbulent medium – the atmosphere/air – in and through which such relations play out.

Again, in the context of cycling, relations do not just take place between mobile bodies, transport technologies, and transport infrastructures. Relations also take place within the context of the specific 'natural' environment moved through (Simpson 2019). The 'weatherworld' moved through affects cyclists and produces specific felt experience of movement (Ingold 2007). For example, cycling into headwinds can affect cyclists in reducing their capacities to make swift progress, and crosswinds can destabilize and disconcert given the way cyclists are potentially pushed into the path of others. In this sense, "the weather in which one stands can be as much responsible for generating a sense and use of place as the ground on which one stands" (Pillat cited in Veale *et al.* 2014: 26). In being amid this weather-world, "Weather constantly makes and remakes place" (Vannini *et al.* 2012: 364) and the subjectivities that inhabit them. Everyday phenomena like wind and rain can affect bodies in their inhabitation of and movement through a host of spaces, shaping capacities to act (Vannini *et al.* 2012).

Staging atmospheres

An important point to recognize here in such circumstances of interaction is that atmospheres do not just occur organically through the free relation of bodies and things. Rather, atmospheres are very often 'staged', 'engineered', or 'designed' in a range of ways (see Bille *et al.* 2015; Edensor and Sumartojo 2015; McCormack 2008a). For example, Bohme (2017) draws attention to the importance of set-design and other broadly architectural practices which might 'set the stage' for a certain atmosphere to emerge. Here, the aim is to produce a certain 'climate' on stage meaning that atmospheres are seen to be something that can be produced through the management of various environmental parameters (lighting, sound, the arrangement of furniture, etc.). This has, though, been considered in wider realms beyond the stage itself (see Adey *et al.* 2013; Thibaud 2015). As Bille *et al.* (2015: 31) suggest, there are a range of "cultural, economic or even political premises that lay the ground for the sensuous and emotional feel of a place" (see Figure 3.3). While it might seem odd that such a vague notion as an 'atmosphere' could be the target of design and other interventions across such areas of social life, geographers have begun to identify a range of efforts to "intentionally shape the experience of, and emotional response to, a place through the material environment" and so uncover agendas that seek "to affect people's moods and guide their behaviour for aesthetic, artistic, utilitarian or commercial reasons" (Bille *et al.* 2015: 33). These 'manipulations' are significant to how spaces come to be experienced but equally "often work in tacit or ambiguous ways, making them easy to overlook as social, economic, and political instruments" (Bille *et al.* 2015: 31).

It is possible then to look at various ways in which attempts are made to shape the 'mood' of spaces through the organization of bodies and objects in such spaces. This is clear in Tim Edensor's (2015) discussion of 'match day atmospheres' in a football stadium in Manchester. Here, Edensor considers the role that fans play in creating such atmospheres but also the specific stadium spaces they come to occupy which, he suggests, 'condition' such atmospheres. Alongside a range of contextual factors – weather, the significance of the

Figure 3.3 Designing 'tranquil' atmospheres? (*Author's own*).

match, the level of rivalry between participating teams, and so on – Edensor (2015: 82) explains the significance of such stadia as follows:

> The football stadium serves as an enclosed theatre in which the sporting drama of the match unfolds, and it tends to house a particularly responsive audience who are themselves part of the drama and can potentially influence what happens on the pitch. The stadium possesses architectonic qualities that promote and contain levels of noise, and organise the distance between fans, and the closeness of fans to the pitch and players. These spatial contexts contribute to atmospheres of varying intensity that continually emerge during a match.

There are, therefore, a range of factors at play in the ongoing co-production of such footballing atmospheres. While co-produced in the interactions of stadium space, players, ball, fans, referees, and so on (see Latham and McCormack 2004), that is not to forget that positive atmospheres can become targeted (or lost) through processes of design. Edensor draws specific attention to how different spaces, such as stadia, can hold certain affordances and so affective potentials by highlighting changes in their design. Contrasting Main Road and the Etihad Stadium – the past and present stadia used by the English team Manchester City – Edensor suggests that the former held 'thicker' atmospheres than the latter given the strong sense of belonging that fans had accumulated over time and through repeat attendance. Important also is the seating style of the stadium given the very different experience that might emerge when standing on vast terraces compared with contemporary seated arrangements. When it comes to the new stadium, while efforts are made to (re)generate more familiar and positive affects – through the production of pre-game social spaces, the playing of music in the stadium, amongst other things – this is potentially felt to be less affective given the way it mirrors broader trends toward "generic designs" that are "bereft of the accretions and patina of age", and as the stadium is situated in a convenient locations (i.e. accessible via key transport hubs) but equally "cast off [from] the traditional associations of clubs with particular areas of the city" (Edensor 2015: 86).

When it comes to the relationships of a body and an atmosphere, then, it is important not to think solely in terms of an outside-in logic whereby a body arrives and is affected by that atmosphere (Dyson 2009). Rather, arriving bodies will bring something to that atmosphere either modifying it or responding to it based on certain susceptabilities (Bissell 2010). The specific disposition those bodies bring to such encounters matters and so the staging, design, or engineering of an atmosphere is a contingent and potentially provisional achievement.

Mediating affect

If affects constitute a fundamental part of the unfolding of everyday life and the relations that occur between people:

> Understanding how power functions in the early twenty-first century requires that we trace how power operates through affect and how affective life is imbued with relations of power, without reducing affective life to power's effect.
>
> (Anderson 2014: 8)

In light of this, work associated with NRTs has begun to explore how "affective life" becomes "an object-target for specific and multiple forms of power" (Anderson 2014: 4) and so how "affective responses can be designed into spaces" (Thrift 2004b: 68). The concern, in many ways, shifts from the sort of definitional debates mentioned earlier (i.e. how it differs from emotion, feeling, etc.) and more to what a focus on affect allows us to become attuned to (Anderson 2014). This is significant given that some initial critiques of work on affect in geography suggested a lack of attention to power relations when it came to different bodies' capacities to affect and be affected (see Tolia-Kelly 2006a). As part of this, there have been a range of attempts to understand how affective life is both actively manipulated for economic ends (Ash 2010a), but also how affect is a fundamental part of contemporary politics in democratic societies (Amin and Thrift 2013). In combination, this has led to a concern with how to understand how, as a "target of intervention", affects constitute a "point where a relation of power meets a form of knowledge" (Anderson 2014: 24). In this, there has been an emerging concern with how affects are mediated by specific material arrangements that come to be rolled up in our day-to-day lives. These meditations are argued to order or re-order social life and the affects that unfold within this.

A key motivator behind NRTs' turn toward affect as a source of both conceptual interest and practical exploration, then, related to the political possibilities, and problems, that affects bring to light. Drawing inspiration from William Connolly's (2002) discussion of 'neuropolitics', Thrift (2004b: 71) suggests that a concern for affect can lead to "a politics which recognises that political concepts and beliefs can never be reduced to 'disembodied tokens of argumentation'". Thrift gives the example here of work concerned with difference and identity which, he argues, previously was focused primarily at the level of discourse, signification, and representation. Here, the differences between bodies are identified on the basis of variations from an understood norm or majority position or from various minorities that might be present in a given situation. In response, Thrift (2004b: 71 [citing Connolly]) suggests that difference and identity *also* operate on other registers, including the affective, whereby:

> surpluses, traces, noises, and charges in and around the beliefs of embodied agents express proto-thoughts and judgements too crude to be conceptualised in a refined way but still intensive and effective enough to make a difference to the selective way judgements are formed, porous arguments are received, and alternatives are weighted.

Turning to affect then presents us with "the idea of a politics aimed at some of the registers of thought that have been heretofore neglected by critical thinkers" (Thrift 2004b: 71). That neglect is concerning given the ways in which a host of political impulses can be emotively charged and become of interest precisely because of the feelings they provoke (Amin and Thrift 2013). For example, Thrift (2008) shows the significance of affect to political campaigning given the ways in which this has increasingly drawn on practices of marketing that seek to catch attention and generate/sustain engagement (see Box 3.2). This includes the increasingly diffuse mediatization of campaigning (both through mainstream and social media), political actors increasingly becoming stylized commodities to be sold so as to bypass the need for sustained engagement with actual political issues, and the increasing extent of campaign marketing practices and the targeting of the campaigns suggested above toward particularly communities' affective susceptibilities. There has been, particularly in the USA, the development of a host of "sophisticated technologies to both track and work on public emotions and to build political content and direction, often guaranteeing success to the political forces that are most able to chip away at the hyphenated joint between thought and feeling" (Amin and Thrift 2013: 157).

BOX 3.2 THE TRUMP AFFECT

As part of NRTs' concerns around the politics of affect, an important theme has been the understanding of politics as increasingly taking place via techniques of "swaying constituencies through the use of affective cues and appeals which are often founded in spatial arrangement" (Thrift 2008: 248). Such points have been explored through reflection on a range of political figures and their campaigns (see Berlant 2007; Massumi 2002). This can be clearly seen in what Anderson (2017) calls the 'affective styles' of Donald Trump's campaigns, first for the Republican presidential nomination and subsequently for the presidency itself.

Trump's campaigns came to resonate in certain ways with the broader 'affective conditions' of post-financial crash, post-9/11 America. Given his background as a media personality as much as a business figure, it is not hard to see a range of strategies employed by Trump which draw on his 'media brand' and the sorts of approaches that succeed in that context, namely that controversy entertains and entrains more than basic competency. However, beyond personality and being media-savvy, Anderson (2017) draws attention to how Trump's approach came to work specifically within the context of certain affective conditions. Trump's style resonated and came to fit within "spreading economic precarity" and "the irruption of populisms of the left and right across Western Europe and North-America ... in which an anti-politics mood intensifies as it is expressed in forms of cynicism, contempt and apathy" (Anderson 2017: n.p). In this sense, Anderson (2017: n.p. [citing Connolly]) suggested that "We might think of Trump as a 'persona' who acts as both a 'shimmering point' and 'catalysing agent' for a set of political feelings and conditions that extend beyond him".

There are a number of features of Trump's affective style, but one of the most prominent is perhaps the constant dualism of winning and losing or winners and losers in which Trump positions himself exaggeratedly as a winner or the winner. As Anderson (2017) notes, Trump presented losing as a 'generalized condition' of the United States throughout his campaigning. In presenting this condition, Trump taps into common feelings held amongst a host of voters of having lost out to a range of (largely external) forces and a range of villainous figures that represent such forces and/or are responsible for such loses – immigrants (bad hombres), foreigners (terrorists), the media (fake news), 'mainstream' politicians (crooked), and so on. In contrast, Trump boasts of his success and his winning. He revels in it in ways which might be thought of by many as crass – boasting about how much money he has made, for example. But through this he offers "renewed promise and possibility of sovereign action" (Anderson 2017: n.p.), of taking charge and winning. Through a combination of dramatized entertainment in his speeches (mock impressions of 'losers' with whining voices) and arrogant exaggerations over his own prowess and success (at times presented in misogynistic, sexualized ways), Trump gives these voters "permission to enjoy their resentments and grievances, permission to enjoy hate" (Anderson 2017: n.p.). He captures the imagination and votes of various individuals by appealing to them at a particularly affective level of concern; it's not about the facts, it's about the 'feels' that resonate.

This concern for a politics of affect also presents opportunities if it is aligned with a specific disposition tied to experimentalism and a concern for how "world-making capacities" might be mobilized through the production and propagation of certain affects or atmospheres (Amin and Thrift 2013: 4). It becomes a case of seeing "politics as the art of generating affective fields" through the "active cultivation of alternative feelings" (Amin and Thrift 2013: 70, 158). Or as Stewart (2007: 3) suggests, affects' "significance lies in the intensities they build and in what thoughts and feelings they make possible". But, at the same time, it also makes us aware of how "those in power have turned to these registers as a fertile new field of persuasion and manipulation" (Thrift 2004b: 71). Such a realization draws attention to how much of day-to-day life takes places within a "marketplace of affects, filled with values, images, analogies, stirring narratives, and moral sentiments" which variously vie for and shape our attentions (Amin and Thrift 2013: 48). As Barnett (2008: 187 [citing Thrift]) suggests, this means that for NRTs, "Understanding affect is a pressing political task ... because 'the systematic manipulation of "motivational propensities" has become a key political technology'".

It is important, though, to strike a note of caution here regarding the strength of this emphasis on the manipulation of affects through media and other means. For example, Barnett (2008) argues that focusing on a pre-reflective realm of unthought-of action – the autonomy of affects' unfolding – "can inadvertently reestablish discourses that position consumers of media as passive dupes who are shaped unwillingly by the technologies with which they engage" (Ash 2010a: 655). Here, there has been an emphasis on how political subjects are manipulated without them actually knowing about it. Barnett (2008: 193) characterizes this as 'hypodermic' model of power whereby 'the media' is ascribed "the ability to inject their preferred message into the minds of their audiences". This gives such means of manipulation a "remarkable determinative power in infusing affective dispositions under the skin of their audiences" (Barnett 2008: 193). This leaves it very hard to propose any affirmative account of the politics of affect given its implication that this itself might end up being engaged tactically in such manipulations given their preference for open/creative ethical stances (also see Amin and Thrift 2013).

In exploring such issues of affective manipulation, various forms of consumer technology have been considered by geographers in terms of how they intervene into affective experiences, influence moods, and attempt to both amplify and dull certain affective responses and the feelings associated with them (see Ash 2010a). There has been concern for how this affective "realm ... is increasingly susceptible to new and sometimes threatening knowledges and technologies that operate upon it in ways that produce effective outcomes, even when the exact reasons may be opaque" (Thrift 2004b: 71). But at the same time, heeding Barnett's (2008) concerns, there has also increasingly been an explicit awareness of the limited effectivity of these interventions and so the fragile and contingent nature of such outcomes when it comes to affective 'manipulation'.

Such affective manipulations via technology can be seen in a host of contexts. This can unfold as part of the background of our day-to-day lives in something as banal as how places might (be made to) smell or sound and how these smells/sounds might be actively shaped to produce certain affects and feelings (Anderson 2014). One case where we can clearly see this two-sided situation of contingency and mediation comes in James Ash's (2010a) discussion of video game design and the sorts of 'worlds' these design practices produce. There is a great deal of affective manipulation that can take place in the designing of video games. Much of what is possible for the gamer is determined by those designing the game. Using both overt and subtle techniques, "game designers are able to shape the spatial and temporal experiences of videogamers and their practices by manipulating the rules of videogames (albeit imperfectly)" (Ash and Gallacher 2011: 360). Game designers not only act as the architects of the spatio-temporal confines in which game play takes place – designing the specific spaces interacted with and the timeframes on which this happens – but can also significantly alter the experiences and affects of the gamer by, for example, varying the responsiveness of controls, experimenting with the specific capacities and reactiveness of game avatars/characters, and so on. The ultimate aim of this is the proliferation of positive affects and limiting negative ones. This means proliferating affects that capture and maintain attention and so make the gamer want to play more and mitigating against affects emerging from poor game design which might lead to frustration or, ultimately, the gamer becoming disenchanted with the game (Ash 2010a).

Here, Ash (2010a: 655) makes it clear how care needs to be taken so as to not ascribe too much efficacy to such designing practices as

> affective manipulation is necessarily a fragile achievement that is prone to failure and always reliant upon being continually reworked in the creative responses users develop in relation to the designed environments with which they interact.

While gamers' actions can be shaped by the rules of the game, in many games, there is also some degree of contingency to the outcomes of the relations of gamer and the game's architecture. This means that the relationship between gamer and designed game environment is

> a complex, problematic, and ongoing struggle between the openness and performative play of contingency and chance (which

emerge through the techniques and intelligences that users develop as they become skilled at these games) and the mechanical systems and calculative rationalities through which these environments are designed.

(Ash 2010a: 655)

Designers – be that of games, urban spaces, or political campaigns – are not all-powerful here, and their intentions might not come to fruition as a result of the unintended responses of the constituent body-subjects involved in a given set of circumstances.

Conclusion

The affective moment has passed in that it is no longer enough to observe that affect is important: in that sense at least we are in the moment after the affective moment … affect has simply become an accepted background to so much work, a necessary part of the firmament through which the forms and shifts of any analysis are extruded.

(Thrift 2010: 289–290).

This chapter has introduced affect as both a key idea for NRTs but also as something which has constituted a key concern for empirical research inspired by NRTs. This has included an attempt to define some of the key features of NRTs' understandings of affect, how such affects have been shown to animate a host of everyday settings, how affects might be experienced collectively and communicated within collectives, and how all of this might be mediated and manipulated by various forms of media and technology. Along the way, critical points important to NRTs' engagement with affect have been touched upon – its argued masculinist undertones, questions of difference in capacities to affect/be affected, and how a politics of manipulation might leave open and understand questions of alternative political action/response. Importantly, here, such critiques have affected those working with NRTs and the accounts of affect they provide. This body of work has been shown itself to be an evolving thing which has both refined and expanded its accounts of social life in light of these tensions (either by way of direct response or implicit recognition).

While affect has come to be bound up with both the development of NRTs and its more contentious contribution to the conduct of cultural geography, affect has come to be accepted as a significant feature of the geographies of everyday experiences of space and place. A glance at the contents pages of a range of mainstream human geography journals will regularly show a range of papers with affect in the title or as a keyword. From 2008 to 2018, over 120 articles were published in the journals *Transactions of the Institute of British Geographers*, *Environment and Planning A* and *D*, *Social and Cultural Geography*, and *Cultural Geographies* alone with 'affect' as a keyword. And that doesn't necessarily capture those articles where affect is something engaged with as part of a broader account of individuals' or collectives' experiences in relation to a host of topics. At the same time, though, such affective experiences remain contentious based on the ways in which they are (or aren't) understood to intersect with questions of social difference and power. There is, then, an ongoing concern and a need to engage with how affects are both experienced by specific individuals and amongst collectives, with how affects take place as part of a range of pre-reflective actions, but equally come to be the target of intervention and shaping based on our interactions with a host of human and non-human materialities and technologies.

Further reading

Anderson (2014) presents one of the most sustained engagements with affect thus far published in geography. While approaching a range of conceptual themes and so being quite advanced in level – these themes particularly relate to how affects mediate and are mediated in various arenas of social and political life – the points made are grounded in a range of empirical contexts and everyday settings. **Dewsbury (2009)** provides a relatively advanced overview of work on affect and particularly the conceptual starting points from which this has emerged. This considers the different ways that affect has been approached – as something tied to the practices of different bodies, as a pre-reflective force circulating in and between bodies, and as something different to related terms like emotion. There is also discussion of the 'politics of affect' and the sorts of ethos on which that might be based. **Bille et al. (2015)** provide a clear overview of work interested in atmospheres. This covers work in geography but also related discipline like anthropology and philosophy. In addition to background/definitional questions, this piece also considers how atmospheres are 'staged' through aesthetics and design and how this has implications for how spaces are experienced. In this annotated bibliography, **O'Grady (2018)** provides a general overview of work on affect, situating this in broader trends in geography and in relation to a range of key themes. In particular, a range of references are briefly summarized/suggested covering the definition of affect and a host of related themes that have featured prominently in geography's engagement with affect. This would present a very useful resource in finding further sources of reading on affect.

4

NON-REPRESENTATIONAL
THEORIES AND MATERIALITY

Introduction

Rethinking the way that geographers understand and research materiality has been a longstanding concern for NRTs (see Thrift 1996). NRTs have focused in various ways on how bodies are enrolled in relations with a whole host of 'stuff' that is made use of in the unfolding of action but also, in themselves, hold capacities to affect those human bodies (and, for that matter, other non-human bodies). Rather than viewing the non-human world as an inert background to the human world, NRTs seek to find means of giving (a more) equal weighting to both humans and things (Anderson and Wylie 2009). NRTs' starting point is that the world is made up of a whole host of things thrown together and that the constant (re)assembling of such things in relations, and the maintenance of such networks of relations, is what keeps that world going.

This concern for materiality is by no means unique to NRTs. Over the course of the last 20 years, there have been calls for geographers to 'rematerialize' human geography (see Jackson 2000; Lees 2002; Philo 2000). Further, there has been a range of work that has moved in parallel to NRTs which has considered materiality in a range of ways. This has included a diverse range of 'material cultures' literature (Cook and Crang 1996; Desilvey 2006; Edensor 2005; Gregson 2007; Tolia-Kelly 2004), work interested in the fleshy matters of bodies (Colls 2007; Fannin 2011; Longhurst 2001), and a host of more-than human or hybrid geographies of nature, science, and technology (Bingham 2006; Hinchliffe 2007; Kinsley 2014; Roe 2006; Whatmore 2002).

This chapter will start by looking at the questions various geographers have asked about cultural geography's attention to the material world and, in particular, the various calls made to 'rematerialize' human geography. This will include a discussion of NRTs' responses to such concerns. In light of NRTs' reimaginings of what materiality is, the chapter then turns to various questionings of state, scale, and materialization which are important to NRTs' accounts of social life. Taking this questioning of the material/immaterial further, the chapter concludes by looking at a key arena where geographies inspired by NRTs have explored such materializations: technical objects.

What do we mean by 'matter'?

There is a diverse and longstanding history of cultural geographers engaging with various materialities in their research. For example, the Berkeley School of Cultural Geography focused on the morphology of the landscape and the diffusion of culture through space/time through attention to cultural artifacts and vernacular building styles (see Kniffen 1965 and Chapter 1). This was clearly concerned with materiality. There are also the geographical–historical materialisms that emerged through, and in light of, David Harvey's readings of the work of Karl Marx. These are concerned with the social and productive relations that exist behind human subsistence (and relative prosperity) and their variance/evolution over time and space (see Harvey 1982). While concerned with what might be seen to be 'immaterial' structures, such work was also very concerned with their material effects upon people's well-being (Philo 2000). However, with the evolution of New Cultural Geography in the latter part of the 20th century, where the focus fell increasingly on 'immaterial' forms of culture (representation, discourse, and text (see Chapter 1)), such material concerns were not so immediately evident (Latham and McCormack 2004). Combined with its critiques of the 'object fetishism' of the Berkeley School (see Duncan 1990), this apparent emphasis on the 'immaterial' led some geographers to argue that New Cultural Geography's

> obsession with meaning, identity, representation and ideology was in danger of replacing studies that were more firmly grounded in material culture or concerned with socially significant differences of gender, class, race, sexuality or (dis)ability.
>
> (Jackson 2000: 9)

Such concerns led to a series of calls at the turn of the 21st century for geographers to 'rematerialize' social and cultural geography.

Rematerializing cultural geographies?

Recent calls for a rematerialized cultural geography have taken a variety of directions. For example, Jackson (2000) called for greater attention to 'material cultures' and so how material form and social significance combine and 'make a difference'. Philo (2000: 33) argued that New Cultural Geography's immaterial focus meant a lack of attention to more obviously "bump-into-able, stubbornly there-in-the world kinds of 'matter'", and so a need for attention to the materiality of the social spaces people actually live out their social lives within and the actual bodies that do that living. Lees (2002: 101) appeals specifically to 'new' urban geography as a body of work that could offer inspiration when it comes to rematerializing human geography through linking the material and the immaterial given that such work remained "more firmly grounded in material culture or concern with socially significant differences".

These critiques of 'New Cultural Geography' have been questioned, and arguments have been made both that this work simply did engage matter (just new materials for geography) and in terms of its concerns for the sorts of material exclusions experienced by specific groups (see Anderson and Tolia-Kelly 2004). Part of this response revolves around a tension in these calls to 'rematerialize'. Materializing is often equated to a form of grounding investigations in some kind of tangible 'gritty' reality and, in turn, a step back from more abstract forms of thinking and analysis (Latham and McCormack 2004). This comes from the different conceptual starting points present when it comes to geographic research on materiality (Kirch 2013). As Lees (2002: 102) notes, part of the problem here is that within the:

> call to rematerialize geography ... geographers tend to use the material and immaterial as a shorthand for tensions between empirical and theoretical, applied and academic, concrete and abstract, reality and representation, quantitative and qualitative, objective and subjective, political economy and cultural studies, and so on.

However, matter means different things and is deployed in reference to different things across the work discussed above. There are a host of different 'materialisms' at play here, and the immaterial is not always necessarily about the theoretical, representational, metaphysical, or transcendent (compare McCormack 2012 and Kirch 2013).

Reimagining matter

Rather than seeking to 'ground' geography in considerations of matter and the material – or to 'rematerialize' at all – NRTs have drawn on a series of recent materialist ideas which question what we mean by the terms material and matter themselves. The argument that follows from this is that geography does not need to be rematerialized *per se*, but rather that geography needs to be clearer about the terms being employed and how matter, materiality, and immateriality are understood. This has taken place as part of a broader reassessment of materiality in recent work in geography and the various "inventive re-imaginings of 'the materiality of matter'" taking place therein (Anderson and Tolia-Kelly 2004: 669). Here, matter often appears as something "too unruly ... to simply be 'included' at the expense, or in addition, to a focus on 'culture'" (Anderson and Tolia-Kelly 2004: 670). Matter is not a stable starting point here. As Anderson and Tolia-Kelly (2004: 669) continue:

> This renewed sensitivity has aimed to demonstrate how the qualities that mark space-time, and bind space-time into wider sets of relations, change according to the processual movements of matter.

This sensitivity can be seen clearly in Latham and McCormack's (2004) rethinking of materiality. Here, they argue that matter should not be invoked as a 'reality check' against theoretical abstraction. Instead, the relationship between 'the material' and 'the immaterial' is seen to be a key question for contemporary human geography. For Latham and McCormack (2004: 703), this relationship is not to be understood as one of opposition but rather one where the force of the immaterial gives materiality "its expressive life and liveliness independent of the human subject". This suggests that geography has not become detached from the material reality of living through its attention to the immaterial elements of cultural life. Rather, it is the case that geographers have not "engaged with sufficient conceptual complexity" with the excessive nature of materiality (Latham and McCormack 2004: 704). In some sense, the argument here is that geographers haven't been abstract enough in their engagements with materiality (also see McCormack 2012). As they continue:

> It is not enough to use the 'material' and 'materiality' in such a way as to invoke a realm of reassuringly tangible or graspable objects defined against a category of events and processes that apparently lack 'concreteness'. Rather, we only begin to properly grasp the complex realities of apparently stable objects by taking seriously the fact that these realities are always held together and animated by processes excessive of form and position.
>
> (Latham and McCormack 2004: 704–705)

The key point here is that matter is constantly in process, constantly 'mattering' or 'materializing'.

One way we can illustrate this is in terms of timescale and how two different material objects are in process but in ways that are more or less easily observed. Take two 'material' objects – an ice lolly (or 'popsicle') and a Victorian-era terraced house. Both of these objects are in process, even if that might appear more or less hard to accept.

On a hot day, the ice lolly will melt quickly; the lolly will change state from (near-)solid to liquid as a result of the conduction of heat from the warm air into the icy substance. During this time, the frozen water molecules present will become excited or agitated, move around quicker and quicker, and will loosen their bonds between each other. This will happen in a matter of minutes and potentially faster than it can be consumed by the child holding it. It's easily perceived in that children's hands and clothing and the floor below quickly become sticky and stained with the sweet liquid, commonly experienced in economies with access to such products and the means to store them, and so banal as to seem patently obvious.

However, the house is also potentially falling apart, we just (hopefully!) cannot see it happening in front of us as we look at it (Dewsbury 2000). The house in question is built from bricks. These bricks themselves are objects that have emerged through the interaction of a series of materials and processes (Latham and McCormack 2004) and have been assembled together into a specific form held together by mortar. However, the bricks and their arrangement into the shape we know as the house are not static. Take the un-rendered chimney stack at the rear of the property. Over the course of a century of wind and rain, the mortar holding its bricks together has been eroded. In turn, this time over years (perhaps decades), moisture has seeped through the resultant gaps and, under the force of gravity, has made its way down into the loft space and timbers of the roof adjoining the chimney stack. These timbers have rotted, and the mortar on the internal wall beneath the stack has turned back to sand. These processes of materialization go on without the occupants becoming aware of them; they are sat in an office underneath it day after day marking essays, writing lectures, writing this book, and so on. When they do venture into the loft on occasion (to retrieve a guitar case, for example), the signs of change are not clearly visible under the limited illumination of a single low-wattage lightbulb. It's only when a final process of materialization occurs – the dampening of ceiling plaster and the sporing of mold – that these matters become the subject of closer investigation. Said investigation results in a decision to have the chimney stack removed and the roof and its timbers replaced. And that process itself leads to the realization that the old roof tiles are made of asbestos, meaning a host of other materializations have potentially been set off as the force of tools fell upon them.

Looking at things this way, the concern becomes with relative consistencies in processes of interaction rather than concreteness. It means that:

> to argue for the importance of materiality is in fact an argument
> for apprehending different relations and durations of movement,
> speed and slowness rather than a greater consideration of objects.
> (Latham and McCormack 2004: 705).

What we end up with is an attempt to "hold onto the relational and emergent imperatives of material forces" which lie behind things that are always in a process of being made, that are part of a "co-fabrication of social-material worlds" (Whatmore 2006: 603–604).

Multiplying material imaginations

Pursuing such a concern with processes of materialization, a range of work associated with NRTs has sought to think past notions of the material as being simply solid or concrete and so to consider other less solid but nonetheless 'real' materialities and the immaterial processes that unfold within such materializations. Again, this is not about grounding geography in the solid or the immediately tangible. Rather, it is a question of material geographies trying to "address its elemental prejudices" which tend toward "phenomenal stuffness" and to recognize that "If we begin with the assumption that the elemental is the material as solid, then we limit drastically the scope and relevance of our conceptual ecologies" (Jackson and Fannin 2011: 435 & 439). There are a range of conceptual questions here that are significant in terms of both how we understand matter itself and how we come to research it. There are also a range of social and political questions in terms of how such matters are significant to a range of contemporary ecological issues – anything from urban air quality to chemical warfare to renewable energy production (Engelmann and McCormack 2018).

As part of such moves, Anderson and Wylie (2009) suggest that to be 'thoroughly materialist' we must multiply our 'material imagination' so that matter can be understood and attended to according to the principle that matter takes place with the capacities and properties of any state (solid, liquid, gas) or element (earth, wind, fire, etc.). From this, there has been a growing body of work on such 'elemental geographies' (see Adey 2015; Engelmann 2015a, 2015b; Jackson and Fannin 2011; Martin 2011; McCormack 2015b). This is clear in its attempt to think through materiality by moving in directions that open up a very different vision of what matter is and how it comes to matter.

We again encounter calls here for an attentiveness to "the complexity, and indeterminacy, of matter and ... [to] how qualities of liveness are internal to, rather than in supplement or opposition to, the taking place of matter and materiality" (Anderson and Wylie 2009: 319). This emphasizes the dynamism of materialities in that materiality here is something that is assembled amongst a multiplicity of relations, maintained as a host of 'turbulences', and with that is open to change. There is little reassuring in such materialities. Despite any appearance of solidity or obduracy, they could, at any moment, fall apart or change as they enter into different relations. Materiality here is a process held in tension between order and disorder, a sort of intermediary state. In this, subjectivities and worlds *devolve* and *precipitate* from such materialities (Anderson and Wylie 2009). What we have are sensible relations with the world where body and world, body and materiality are entwined and interrelate.

BOX 4.1 ELEMENTAL AESTHETICS

When it comes to concerns with the elemental, the focus thus far has commonly been on air and other elemental circumstances in political terms. Elemental geographies unfold in the context of megacities, the governance of air and volume and the breaching of geopolitical demarcations, and so on (see Adey 2013; Adey *et al.* 2011). In expanding such elemental geographies, Engelmann (including work with McCormack) seeks to explore more aesthetic (but no less political) ground and

> how the agency of atmospheric milieus participate in the crafting of new forms of association and life, not least because in sensing variations in these milieus we are also sensing the force of something the origin of which lies beyond our sphere of planetary concern.
>
> (Engelmann and McCormack 2018: 243;
> see Engelmann 2015a, 2015b)

The sort of "elemental aesthetics" which they are concerned with is "organized around the challenge of making sensible through various experiments how the elemental shapes and sustains diverse worlds" (Engelmann and McCormack 2018: 442). This elemental aesthetics comes through clearly in Engelmann's (2015b) discussion of Dryden Goodwin's artwork 'Breathe' (on solar energy and air, see Engelmann and McCormack 2018). The artwork took the form of 1300 hand-drawn sketches of a young boy breathing which were digitally processed and sequenced into a 12-minute animation. This was projected onto the side of Gassiot House (by the River Thames, London) and looped continuously from sunset until 1 am throughout October 2012. The location was significant in that it was both proximate to St Thomas Hospital (the animation could be seen from its entrance) and across the river from the Houses of Parliament.

In developing a 'poetics of air', Engelmann's (2015b: 430) discussion of 'Breathe' develops "an awareness of the simultaneous material, affective and aesthetic impressions of air". Emerging from the context of discussions between scientific researchers working on the matter of air quality in London and its impacts on children, 'Breathe', for Engelmann (2015b: 432)

> expands our material optics and attentions from the solid to the airy, and mobilises the figure of 'the breather' as metonym for a collective being-in and witnessing air ... [in the way that it] dissolves distinctions of body-environment boundaries, renders explicit air's materiality and fosters an openness to the affective intensity of air in shaping the patterns of atmospheric space-time.

Interesting here is Engelmann's (2015b) focus on the relationship between the production of the artwork and the sense of air's materiality that this brought. Engelmann discusses this in terms of the 'sequencing' of breath and how this meant that air was 'spaced-in' to the artwork for viewers. This comes from the fact that each of the 1300 animated frames from the animation were completed by hand – bringing with it its own sense of texture and surface – and the way in which these images sought to demonstrate the minute changes that take place when air permeates the human body, giving air a sense of form or solidity. A multitude of variations occur here when breathing and can be seen in the changes in rhythm and speed which unfold in the (normally) reflex and unthought action of breathing. In this:

> *Breathe* offers a performance of air-body relations that actively registers the passage of air through the atmospheres inside and outside the body, and does so through ... Cultivating distinct experiences of bodily permeability, air's substantiations and the affective intensities born by airy impressions.

> (Engelmann 2015b: 440)

What we have here then is a disposition to the material whereby there is an "ongoing process of learning to become affected by the force of the elemental through experiments with a range of crafts, materials, and devices as they are moved, pulled, and pushed by the elemental" and so a disposition which "foreground[s] the immersive elemental conditions in which forms of life take place" (Engelmann and McCormack 2018: 255).

One area of work where we can see this sort of concern for turbulent ma-terialities which take place with varied capacities and properties, in multiple states, and in relation to a host of (non-)human bodies, is in discussions of various 'aerographies'. Here, geographers have asked: "What if we began with a different element, one other than the earthy crust usually the focus of geographical thought?" (Jackson and Fannin 2011: 438) (see Box 4.1). This work explores how a host of not apparently solid matters circulate and envelop a host of bodies and objects (see Chapter 3). It explores our existence that plays out amid a turbulent medium within which we are immersed and which we rely on to live (Engelmann and McCormack 2018; Ingold 2007). This then leads to an appreciation of:

> the contemporary importance of our immersion within materially complex and fragile balances of earth, air, water, and fire, and the importance of their delicate balance for our specifically human continuity and flourishing.
>
> (Jackson and Fannin 2011: 436)

(Re)thinking matter and materialization

NRTs' interests in matter and materiality have unfolded in light of a range of conceptual framings, thematic orientations, and units of concern. This section will explore several of these developing conversations with work taking place beyond the confines of geography around various forms of 'new' or 'relational' materialism. Amongst such work, we can see a com-mon thread in terms of how materiality takes place in ongoing processes of relation and materialization. This presents a somewhat expanded material realm that takes place at varying scales and along a variety of more or less stable temporalities. That said, the section will conclude by considering the non-relationality of certain objects and how objects might exist and take place independent of human presence (both spatially and temporally).

Assembling actants and networks

A key influence in NRTs' engagements with questions of materiality is Actor-Network Theory (ANT) given its attention to how things – human and non-human – come together in "the heterogeneous genesis of materialities" (Anderson and Wylie 2009: 320; see Chapter 1). ANT can be considered to be a sort of 'materialist semiotics' which is concerned with how order and meaning are built but where that building isn't just made up of signs (for ex-ample, words, images, and diagrams), but also bodies, machines, situations, and so on that are themselves networks (Bingham 1996). In this, there is a "constant hum of the world as the different elements of it are brought into relation with one another, often in new styles and unconsidered combinations" (Bingham and Thrift 2000: 281).

ANT has offered NRTs "a 'new classification of things' (Latour) in which the bounds between subject and object become less easily drawn" (Thrift cited in Roe 2010: 261). This reclassification also brings with it an overlap between NRTs and a range of developments in human geography around more-than human geographies (see Greenhough 2010; Hinchliffe 2010; Roe 2010). For such work, ANT's language of actors (or actants) and networks was seen to "provide a means of navigating those dualisms, such as nature/society, action/structure and local/global, that have afflicted so much geographical work" until the 1990s at least (Murdock 1998: 357). Here, it is shown that the world is much messier and that there are many more hybrid 'quasi-objects' (Bingham 1996).

A key feature of ANT for NRTs has been that ANT seeks to draw attention to how society or 'the social' is not some fixed or stable matter of fact to be used to explain why certain things did (or do not) happen. Rather, society or 'the social' are performatives; they are the outcome of a host of interactions taking place between a wide range of 'things' (Bingham 1996). Matter – both human and non-human, animate and inanimate – is fundamental to the ongoing process of social formation (Thrift 1996); for ANT, it is the "mixing of human actions and non-human materials which allows networks to both endure beyond the present and remain stable across space" (Murdock 1998: 360; also see Hitchings 2003). ANT, then, presents one particularly influential take on the relational understanding of space (see Massey 2005 and Box 4.2).

BOX 4.2 ASSEMBLAGES OR NETWORKS?

In addition to ANT, other ideas have informed NRTs which, at first glance at least, present a very similar constructivist relational materialism. This includes discussions of 'assemblage' or 'assemblage theory' (see Anderson and McFarlane 2011; Greenhough 2011; Müller and Schurr 2016; Robbins and Marks 2009). Commonly, 'assemblage' as a concept is drawn from the writings of Deleuze and Guattari (see Deleuze and Guattari 2004; Dewsbury 2011b) and reference to 'assemblage theory' drawn from the writings of DeLanda (see Anderson, et al. 2012; DeLanda 2006). For Anderson and McFarlane (2011: 124), assemblage

(Continued)

is often used to emphasise emergence, multiplicity and indeterminacy, and connects to a wider redefinition of the socio-spatial in terms of the composition of diverse elements into some form of provisional socio-spatial formation. To be more precise, assemblages are composed of heterogeneous elements that may be human and non-human, organic and inorganic, technical and natural. In broad terms, assemblage is, then, part of a more general reconstitution of the social that seeks to blur divisions of social-material, near-far and structure-agency.

Assemblage very clearly resonates with the sort of attention to (non-human) materiality found in ANT. There is a clear inclusion of human and non-human elements in its frame of reference, and their interactions are deemed to 'compose' social-spatial arrangements through ongoing processes. That is all provisional and maintained through various practices (Anderson et al. 2012), and 'the social' is not used as some kind of pre-existent explanatory tool. There is again an attention to "how common worlds have '[t]o be built from utterly heterogeneous parts that will never make a whole, but at best a fragile, revisable and diverse composite material'" (Anderson and McFarlane 2011: 125 [citing Latour]). In such dis-assembling, agency is again distributed throughout. What is arguably added to ANT, though, is more of an accommodation of the unexpected, a concern for events, and more attention to the specific corporeal capacities of human bodies to affect and be affected (see Müller and Schurr 2016; Thrift 2000).

Studies of science and technology constituted a key laboratory from within which ANT developed. Here, in being concerned with how science might be understood as a social construction (Bingham and Thrift 2000), "In study after study of science or technology in action, actor-network theorists … focused attention on all the elements – test tubes, organisms, machines, texts, and so on – that are juxtaposed in the building of networks" (Murdock 1998: 360). This means that, in being so juxtaposed, these elements are not clearly social or natural; rather, they are hybrid co-productions (Roe 2010). In this and subsequent engagements amongst geographers, a key methodological principle has been that of 'symmetry'. This means that ANT "works with no prior conception of which materials will act and which will function as simple intermediaries for the actions of others" (Murdock 1998: 367). It is important that the researcher "repudiate all a priori distinctions between classes of actors (natural or social, micro or macro, and so on) and limit themselves to following actors as they tie together heterogeneous associations or networks" (Murdock 1997: 738).

This symmetry is related to one of ANT's more contentious principles: that of a 'flatness' amongst actors/actants. Flatness means that all entities should be considered 'agnostically' rather than in terms of how they might commonly be classified (social, natural, technological, etc.). That might mean, for example, paying as much attention to the importance of slug pellets and gel crystals as to plants, gardens, or the gardeners themselves when we think about practices of domestic gardening (see Hitchings 2003). It might also mean that we are concerned as much with databases and marketing materials and sand and tattoos as we are with elephants and volunteers and zoo design when it comes to wildlife conservation (see Whatmore and Thorne 2000). Or it might mean that we are as concerned with scraps of paper, willows, inky sponges, and water vole footprints and feces as we are with urban development agendas and environmental policy documents when it comes to urban living and urban political ecologies (see Hinchliffe *et al.* 2005).

From such a position of symmetry or flatness, different visions of ethics and politics can emerge. For example, in terms of ethics, Roe (2010: 261) illustrates how such an openness to the presence of non-human agencies can sensitize us "to different kinds of (non-human) processes of matter that generate the materialities we know and sense". In the context of research on 'the meat industry' and ethnographic research in slaughterhouses which traced the combinations of practices of animal welfare advocacy, animals' legal status, scientific evidence around animal sentience, husbandry practices, the pre- and post-slaughter characteristics of meat (i.e. coloring and pH values), and practice slaughtering itself, Roe (2010: 263, 276) makes an argument for "a relational ethic" which is "sensitive to the overlooked subtleties of socio-technical and socio-material arrangements producing contexts where events can take place" and so to "the specific expressions of different kinds of non-human matters". In terms of politics, Hinchliffe (2010: 305) suggests that such attentiveness to how various 'things' "reverberate beyond their conventional boundaries, but do so in ways that are not inherently structured or necessarily pre-determined" allows for a 'material politics' to emerge. In the context of researching 'Habitable cities' and the case of cultivating an urban garden, Hinchliffe (2010) shows the multiplicity present in such a garden whereby there are multiple versions of the garden coming from gardeners' engaged in practices of cultivation, local and national urban regeneration concerns, and funding bodies (not to mention the roles of seeds, plants, insects, computers, rain, documents, soils, micro-organisms, gardening tools, sunshine, etc.), and in which the garden itself is both an actor and enacted. Here, politics becomes about more than just words on pages or representations of the garden. Rather, it comes to include how things are done, could be done differently, and so how such different (potential) enactments come to be struggled over. It is what Hinchliffe (2010: 307) calls an 'ontological politics' that is concerned with "the making of realities, their distributions, their effects and the possibility that things could be improved". That isn't 'anti-representational' as words and representations can and do matter. However, that is not to say that they are necessarily the most important actors when it comes to the taking place of such environments and the making of this world.

Such flatness in ANT has proved to be a matter of concern for some geographers. Questions have been asked over whether or not we should "be concerned that there are still some things distinctively human, and perhaps too some things distinctively nonhuman, which get ignored in the process?" (Laurier and Philo 1999: 1061). There is a concern whether such a flatness might homogenize differences that might exist between different actors. It is important here, though, to be clear that symmetry is a *methodological* principle. In that sense, not all entities present within such actor-networks will have the same agency or will have the same freedom to act in certain ways. It is rather that such agencies or freedoms should not be *assumed* before the act of following these actors and their relations within the actor-networks under consideration. As Murdock (1997: 744) explains:

> It is only when the networks have been established, and roles and identities distributed within them, that a clear-cut difference emerges between 'things out there' and 'humans in here'.

It is also the case that such agencies and freedoms once emerged are not static or everlasting. Rather, they will continue to emerge as those networks and actants co-evolve, as new actants enter into relations with them.

New materializations of 'thing-power'

NRTs' concern with non-human agency and how materiality can come to act as part of the ongoing composition of heterogeneous networks of association forms just part of a broader set of developments in the humanities and social sciences that have come to be collectively referred to as 'new materialisms'. Shared amongst such materialisms is an emphasis on "materialization as a complex, pluralistic, relatively open process" and an "insistence that humans, including theoriests themselves, be recognized as thoroughly immersed within materiality's productive contingencies" (Coole and Frost 2010: 7). Here, there is an emphasis on how "life forms are restless, mobile and constantly changing their relationship to their environment; instead of being some thing, life forms are constantly evolving, constantly becoming, shifting in their composition" (Greenhough 2010: 38). In this sense, "new materialists emphasize the productivity and resilience of matter" (Coole and Frost 2010: 7).

Such conceptual framings have been suggested to provide NRTs with further means for thinking about the sort of concerns raised in the previous section. New materialisms themselves have emerged specifically in response to concern over a range of diverse yet pressing issues, such as "climate change or global capital and population flows, the biotechnological engineering of genetically modified organisms, or the saturation of our intimate and physical lives by digital, wireless, and virtual technologies" (Coole and Frost 2010: 5). In such responses, again, we encounter attention to relationality (and specifically what is referred to as 'intra-action'), entanglement, and differing (Dolphijn and van der Tuin 2012). The world here is seen to 'hang together' (Roberts 2012). These new materialisms can be characterized by their critical stance toward the sorts of modernist dualisms which ANT pushed against; collectively, they form "a particular critique of the Cartesian relation between matter and thought" and propose a particular take on nonhuman agency based around (co-)emergence, vitality, force, process, and affective capacities rather than conceptual or practical mastery (Roberts 2012: 2514).

A key source of inspiration here when it comes to new materialist ideas amongst geographers has been the work of Jane Bennett on 'thing-power' and 'vibrant matter'. In talking of 'thing-power', Bennett (2004: 349) proposes a materialism that

> seeks to promote acknowledgement, respect, and sometimes fear of the materiality of the thing and the articulate ways in which human being and thinghood overlap. It emphasizes those occasions in ordinary life when the us and the it slipslide into each other, for one moral of this materialist tale is that we are also nonhuman and things are vital players in the world.

Again, we have reference to a sort of relational materialism whereby the boundaries between human and nonhuman are significantly blurred. In fact, Bennett suggests that thing-power exists specifically by virtue of such relations, that it is a 'relational effect', that things operate as a result of conjunctions.

To illustrate this "curious ability of inanimate things to animate, to act, to produce effects dramatic and subtle", Bennett (2004: 349–350) provides an account of a specific encounter with 'trash', recounting how:

> On a sunny Tuesday morning, June 4, 2002, in a grate of the storm drain to Chesapeake Bay in front of Sam's Bagels on Cold Spring Lane (which was being repaved), there was
>
> > one large men's black plastic work glove
> > a matted mass of tree pollen pods
> > one dead rat who looked asleep
> > one white plastic bottle cap
> > one smooth stick of wood.

In reflecting on this arrangement, Bennett talks about how these items moved back and forth between being 'trash' and 'thing', between being notable only as the residue of human activity and inactivity (a failure to clean or the incivility of littering, pest control, practices of maintenance, and the like) and matter that demands attention given the life it has in its own right (see Bissell 2009c). It is the latter that Bennett talks about in terms of 'thing-power'. As she continues, "I was struck by the singular materiality of the glove, the rat, the bottle cap – a singularity brought to light by the contingency of their co-presence, by the specific assemblage they formed" (Bennett 2004: 350). It was the specific composition here that came to affect and had this been different a different impression (or lack thereof) might have unfolded. When thinking about the thing-power or vitality of such matter, it is the case that:

> each thing is individuated, but also located within an assemblage – each is shown to be in a relationship with the others, and also with the sunlight and the street, and not simply with me, my vision, or my cultural frame. Here thing-power rises to the surface. In this assemblage, *objects* appear more vividly as *things*, that is, as entities not entirely reducible to the contexts in which (human) subjects set them, never entirely exhausted by their semiotics.
>
> > (Bennett 2004: 350 [emphasis in original])

One of the things that has appealed to geographers in Bennett's account is the way in which it walks a line between attending to the 'vitality' of things and so to "give voice to a less specifically human kind of materiality" but still remains interested in humans (Bennett 2004: 348). It is clear in Bennett's (2010: 10) work that there is an effort to "experience the relationship between persons and materialities more horizontally" and so to "step towards a more ecological sensibility". There are echoes of the actor networks and assemblages discussed in the previous section, and Bennett does draw on such ideas in developing her account of vibrant matter and thing-power. Materiality here "is a rubric that tends to horizontalize the relations between humans, biota and abiota" (Bennett 2010: 112). However, this is a tendency rather than an absolute. Key in this is her reference to more horizontally rather than simply 'horizontally'; this is a 'flatter' ontology rather than a flat one (Roberts 2012). The human is not at the center of the analysis but the implications of such relationality and distributed agencies in composing such humans are a clear concern. There is some maintenance of difference between humans and nonhumans but there is at the same time a recognition of the thingness of humans themselves.

What we have, then, is an account of "the capacity of things... not only to impede or block the will and designs of humans but also to act as quasi agents or forces with trajectories, propensities, or tendencies of their own" (Bennett 2010: viii). In this, there is "an existence to a thing that is irreducible to the thing's imbrications with human subjectivity" (Bennett 2010: 348). However, such agencies or forces do "produce (helpful, harmful) effects in human and other bodies" (Bennett 2010: xii) and so also come to be bound up with the processes of subjectification that proceed from the body's relations with them. Therefore, this attention to the vitality of matter articulates "a vibrant materiality that runs alongside and inside humans" (Bennett 2010: viii).

Geographers have taken up this sort of interrelated, entwined materialism in a host of contexts where materialities circulate in and through relations and impact at an affective, experiential level. For example, this has included considerations of the materialities of sounds and how bodies that listen are affected in their sensuous encounter with the force of various sounds. Here, through the combination of listening bodies, reverberant sounds, and sounding contexts, a variety of affects and effects unfold and various resonant subjects devolve (see Simpson 2009, 2017c). Equally, creativity and processes of artistic production have been complicated in ways that direct attention away from the artist as the instigator of creative practice and the artistic object as a product of that. Instead, creativity becomes something that happens amid a broader array of agencies emerging from the interaction of a host of matters (see Williams 2016). Further, in the context of consumption, geographers have expanded the sorts of approaches to architectural design and the affects it can engineer where there is a focus on the limitation of "certain material capacities in order to encourage more economically desirable choreographies" (Roberts 2012: 2518; see Chapter 3). This leads to a consideration of "the inevitable uncertainties of relations that enroll material capacities previously unforeseen by their architects", something that is deemed to be "an inevitability in a world of lively materials" (Roberts 2012: 2518). Such uncertainty might be brought home in the interruption to the carefully cultivated soundscape of a shop floor presented by a mis-behaving toddler and the reprimand loudly delivered by a fraught, over-caffeinated parent, or in the striking juxtaposition of product displays and shop floor design which (don't) work together in various senses. Common to such work, then, is the attentiveness to various "processes whereby materialities achieve specific capacities and effects. The broad focus is on what matter *does* rather than what its essence *is*" (Anderson and Tolia-Kelly 2004: 672).

While Bennett and others have been very influential in recent ge-ographic discussions of materiality associated with NRTs, some critical concerns have been raised about such approaches to materiality (see Abra-hamsson *et al.* 2015) and specifically how they have been employed by geographers (see Tolia-Kelly 2011). Echoing an approach more squarely aligned with the discussion of ANT above, Abrahamsson *et al.* (2015) have questioned the extent to which a focus on 'thing-power' and the agency of objects can simply state that 'x has agency' without considering a range of related questions that circulate around relationality. This might mean, for example, thinking more carefully about the difference between agency and causality; it might mean that "rather than getting enthusiastic about the liveliness of 'matter itself', it might be more relevant to face the complexities, frictions, intractabilities, and conundrums of 'matter in relation'" (Abrahamsson *et al.* 2015: 13). There may be various 'modes of doing' that matter can become engaged in which are 'distributed achieve-ments' and so are not solely the preserve of individual entities. Further, it has been suggested that geographers taking up Bennett's ideas have missed something of their conceptual underpinnings and the political context of their development. Tolia-Kelly (2011: 153) has argued that such geog-raphies lack reflection, critique, engagement, or evaluation – a political engagement with matter – and so lead toward "a surface recording, a de-scription, a mapping or illustration of materialities within a site or those which are observed" (also see Kirch 2013). These 'surface geographies' are argued to "risk delivering a visual collage of what is observed rather than considered through theories of the material, politics, affects or effects" (Tolia-Kelly 2011: 154). While Tolia-Kelly does not offer examples of such surface geographies, it appears as though we find ourselves here in what is becoming something of a refrain in critiques of NRTs' (argued lack of) politics: the lack of satisfaction found in the different sort of politics NRTs do (or do not) offer or the sorts of problems that they raise as matters of concern (see Chapter 2).

Body materialisms

Alongside these developments, it is also possible to identify a strand of work which Bennett (2004) describes as 'body materialism'. Here, under the influence of phenomenologists like Merleau-Ponty and feminist re-readings of such work by the likes of Irigaray and Grosz, this work has

> examined the micro- and macro-political forces through which the (human) body is, amongst other things, gendered, sexed, pacified, and excited. Body materialism, in other words, reveals how cultural practices shape what is experienced as natural or real ... The point here is that cultural forms are themselves material assemblages that *resist*.
>
> (Bennett 2004: 348)

Echoing the focus of NRTs on processes of social-spatial formation, there is a clear attention here to how bodies are always already positioned within such processes of (de)formation.

For example, Colls (2012) has shown how a conversation between NRTs and the sexual difference theories of Irigaray, Grosz, and Braidotti might both provide feminist geographies with a route into NRTs and allow NRTs to engage with a body of work that it has thus far largely overlooked. This emerges in response to criticisms of NRTs which suggest that they risk reproducing an undifferentiated body-subject in their post-humanist accounts of affect which are not expressly concerned with various categories of social differentiation (age, sex, ethnicity, race, etc.) (see Chapter 3). In response, Colls (2012: 431) suggests that NRTs and theories of sexual difference might work together in developing an understanding of embodied "difference that is not tied to opposition, is not determined by identity, is not subsumed by comparison, but is instead understood as an 'ontological' force". Here, the body-subject would be a "provisional coming together of a range of forces that are material, affectual, temporal, social, political, economic, technological and so on" (Colls 2012: 431).

Colls (2012) draws on the work of Elizabeth Grosz in developing such an account in terms of 'force'. Here:

> Forces... constitute and are constituted through what we come to know as 'a subject'. They operate at a range of scales and intensities. They can pass through and inhabit bodies (metabolism, circulation, ovulation, ejaculation), they are intangible and unknowable and yet are sometimes felt by the body and travel between bodies (fear, hope, love, wonder, hate, confidence) and they are produced

by and active in the constitution of wider social, economic and political processes and structures, for example capitalism, democracy,
deprivation, emancipation and discrimination.

(Colls 2012: 439)

From this understanding of force, we arrive at a position whereby "Female
bodies ... are not understood as the product of a patriarchal culture" but
instead:

> are excessive to hierarchical control and as a realm of affectivity
> they are the 'sites of multiple struggles, ambiguously positioned in
> the reproduction of social habits, requirements, and regulations and
> in all sorts of production of unexpected and unpredictable events'.

(Grosz cited in Colls 2012: 435)

Forces are about difference, differentiation, and differing and so both form
the background to, and are active in, the production of sexually and otherwise differentiated body-subjectivities.

One way that this can be illustrated is through Colls's (2007) work on the
processes through which fat bodies are materialized. Here, it is important to
be clear that Colls seeks to move from an understanding of matter as a passive and pre-existent 'thing' that is acted upon by some external agency that
gives it meaning and stability (often in academic accounts, certain discursive
practices which performatively inscribe the body with certain meanings and
values relative to a more desirable norm or 'natural' body (see Chapter 2)),
to thinking about matter as productive and so participating in its own processes of materialization. Bodies exceed such attempt to fix and frame them
given their spatial and temporal contingencies and their internal dynamism;
matter is not to be interrogated as though it is the end product but rather as
something coming into being.

Translating this into the context of bodily fat, in contemporary western
society fat bodies are often the target of judgment on the basis of medical
claims around health but also in moral and aesthetic terms. Such bodies are
seen to represent negative traits like laziness or a lack of control. For performative accounts of matter, the circulation and citation of such normative
discourses act to normalize such views. However, thinking about these bodies and their materialities differently, Colls looks to provide an alternative
figuration of such bodies which pays greater attention to the dynamism of
that matter itself. This does not necessarily mean the development of a positive account of fat bodies. Rather, it acknowledges that there is more to fat
than its discursive construction. Such an account is developed in terms of
bodies' 'intra-actions'. The use of 'intra-action' rather than interaction here

is important. Talking of such materializations in terms of interaction would risk presupposing the existence in advance of the encounter of independent entities which enter into that encounter. 'Intra-action' "instead insists that it is the relations between subject and object that make up the components of the phenomena" (Colls 2007: 354).

The very physicality of matter matters here. For Colls, bodily fat has a range of intra-active capacities. Fat isn't (normally) leaky given that it is positioned below the skin. It is also ambiguous given that it sits between a status of under our skin but also as a substance in and of itself. In this sense, fat is disruptive; fat isn't so easy to situated, fix, or bound. Fat gathers, folds, and touches itself. Fat can be molded (for example, by clothing), and it can vary across time on the basis of weight loss or gain. Fat has certain capacities to do, it has a sort of force. That might be tied up to its relationship with gravity but it also can have its own momentum and movement. Fat can sway, lift, wobble, jiggle, and so on, in ways that body-mass indexes or other forms of measurement fail to capture. Such capacities to do demonstrate "the multiplicity of bodily matter, materialising differently according to the context and relations it has inside and outside itself" (Colls 2007: 354).

A central concern for such bodily materialism then is difference. More specifically, there is a concern for "the unstoppable, everyday transmogrification of micro-differences" that bodies are caught up within (Saldanha 2010b: 284). Arun Saldanha's (2007, 2010a, 2010b) work here in particular has shown that a key challenge in thinking about this is that there is a double form of difference. We are talking about both *different bodies* and *many bodies* at the same time. To think of identity in this way is different to what might often be understood under the heading of 'identity politics'. Rather than being a 'social construct' – meaning that race or gender, for example, would primarily be understood in terms of ideas and representations that are to be contested/asserted – or as a given from birth, Saldanha (2007: 9) sees identity as:

> a heterogeneous process of differentiation involving the materiality of bodies and spaces... shifting amalgamation of human bodies and their appearance, genetic material, artefacts, landscapes, music, money, language, and states of mind.

On the basis of such 'corporeal interminglings', racial difference, for example, "(like all social relations) is a reality involving the interactions, imaginations, and biologies of human bodies" which are embedded "in extremely heterogeneous nets of things and substances" (Saldanha 2010a: 2412). While delineated identities might and do matter in terms of the unfolding of social life having been produced through various practices and processes of socialization, Saldanha (2010b: 294) suggests that such identities nonetheless "cannot exist without an overabundance of feelings and practices", of material forces, which forms a sort of 'molecular infrastructure' for these collective identities.

Such a concern with difference and materiality is challenging given the ways that such corporeal difference "resists becoming known" and cannot simply be understood "through analysing its capture in language" (Saldanha 2010b: 284). Rather, there needs to be a concern with how differences 'aggregate' into particular collective formations (for example, racial identity) rather than them simply being something that is 'acted upon' individually by various discursive socializations (Saldanha 2010a). Saldanha (2010a: 2410) argues that "If subjectivity emerges through corporeality, do collective identities not too?" And as Saldanha (2010b: 294) goes on to suggest, identities "are never formed by one body alone, and the multiplicities of gut feelings and imaginations are paralleled by the multiplicities of the bodies which they stir". On this basis, "Racial identities − 'black', 'white', 'Asian', 'South Asian' − are the *systematization* of affects, over many millions of phenotypically similar and dissimilar bodies, which together inhabit, carry, and gradually change race" (Saldanha 2010a: 2418). As such, aggregation refers to how affective relations 'add up' and so increase the size of a collective of participants and, at the same time, move up through a hierarchy of scales. Thinking in terms of aggregations means that the key question that geographers need to ask becomes: "How does [such] collective embodiment in a racial formation subsequently constrain and distribute its bodies?" (Saldanha 2010a: 2241).

Interesting amongst such work in the context of the range of materialisms discussed so far is the relationship between a focus on differentiated human bodies and the sorts of non-human material forces they find themselves caught up within. As Colls (2012: 441−442) usefully summarizes, this work has developed:

> an account of difference that is not pre-given, hierarchical or oppositional. It also means deploying an understanding of 'a' subject that is constituted through and by force(s) and that is always contingent on the context in which it emerges. Such forces include those that operate inside and outside of the body, those that are intangible, unknowable and yet sometimes felt and exchanged with other bodies and those wider material forces or processes (social, economic, political) that pass through, around and between bodies. This approach requires an openness to and recognition of the multiplicities, contingencies and virtualities of sexual [and other form of] difference rather than assuming knowledge of what constitutes the deployment of a sexually [or otherwise] differentiated subject in advance.

This is quite different to, for example, ANT whereby the world is flat. In these bodily materialisms the human body, while clearly no longer straight-forwardly human, remains a central concern. It is the case, though, that the realm of 'stuff' that acts upon that body, and through which that body acts, is significantly expanded to recognize the unfolding, dynamic, processual but still fleshy nature of such bodies and the materialities of the contexts they are inescapably enmeshed within.

Objects matter

NRTs' critical engagement with geography's material turns has picked up on work referred to as 'Object Oriented Ontology' (OOO) or 'Object Orientated Philosophy (OOP) which seeks to pay attention to the existence and inter-action of objects (or 'things') outside of the necessary presence of a human subject. These ideas are significant when it comes to how geographers think about matter and how that thinking is framed.

While there are a range of subtle differences between the various ap-proaches discussed so far in this chapter, one guiding thread is an empha-sis on relationality and that relations come before their terms. Such work emphasizes the relations rather than anything that might precede those re-lations. There is a clear emphasis on emergence, process, materialization, and the (ongoing) coming into being of matter. While this has gone a long way toward dispelling approaches to materiality that treat matter as some kind of inert substance or passive medium to be used or given meaning by humans, it does also raise questions about what has been referred to as 'non-relationality' (see Harrison 2008, 2009). As Ash (2013: 23) explains, when there is such a focus on relations, this "creates a situation in which an object only appears in moments of transformation or translation when encountering another entity". This makes it very difficult to provide an account of materiality when it is not actively affecting or being affected by some other materiality. If defined relationally, objects appear to have little to no existence apart from such situations. We encounter questions over whether is it not the case that "objects have a series of qualities that appear depending on what or who is encountering that thing, but also have a non-relational core that makes that thing what it is", even if we never quite see it (Ash 2013: 23).

Such concerns over relationality and the status of objects form key con-cerns for OOO. In general, this work seeks to develop a theory of objects rather than just human access to them (Meehan et al. 2013). There is a concern with how we might "probe the 'world-in-itself'; that strange and unflinching infinity that is indifferent to human thought and discourse" rather than just how they "pattern sociopolitical life" (Shaw 2012: 614). In light of this, OOO argues for a position whereby objects have some kind of existence apart from their relations with other objects and that the relations we see objects enter into never exhaust this 'core' of an object. On that basis:

> we find objects are engines of affect, constantly generating difference, but – and this is a critical conjunction – such activity is never exhaustive because there is 'cryptic reserve' within each object. We thus define an object as an autonomous unit of reality that exists independently of its relations and is the site of a Janus-faced polarisation between its presence and absence. Within such a combinatory formulation, the world becomes a series of negotiations between objects that can only affect each other partially.
>
> (Shaw and Meehan 2013: 218)

Defining 'object' is not straightforward. In broad terms, an object is "a unified 'thing' composed of a multitude of features, which are themselves objects. In short: objects contain other objects, which then become their interior components" (Meehan *et al.* 2014: 61). This leaves open the possibility that we might speculate about what it is to be such a non-human object in the world (Ash 2013). And this "fundamentally denaturalizes our relationship to the bits and pieces of the world, revealing an altogether alien armada of force" (Meehan *et al.* 2013: 3).

A key concern of OOO is with how objects might relate to one another (and to humans) outside of humans' consciousness of this or their intentions, beyond an object's correlation with the human. As Bogost (2012: 6) elaborates:

> OOO puts things at the center of being. We humans are elements, but not the sole elements, of philosophical interest…In contemporary thought, things are usually taken either as the aggregation of even smaller bits (scientific naturalism) or as constructions of human behavior and society (social relativism). OOO steers a path between the two, drawing attention to things at all scales…and pondering their nature and relations with one another as much as with ourselves.

There is, though, here an argument for the autonomy of objects outside of their use by humans and that the impacts that they have do not completely disclose the nature of that object.

To illustrate this set of developments around the existence and autonomy of objects we can take a relatively clichéd philosophical thought experiment and ask the question: if a tree falls in a forest and no one is around to hear it, does it make a sound? Initially, this appears to be a question of perception. Commonly, we would understand sound as vibration registered in and by an ear and recognized as sound by a nervous system. In this sense, the answer would be no – if there are no ears present, there is no sound. However, that is not to say that the falling tree, and the vibration it produces, (a) does not take place or (b) is not registered in some way by animate or inanimate objects. In some ways, to ask if a sound is made is to ask the wrong question. We instead could think of it as a 'geo-event' or "the transformation of a world by 'inexistent objects' and the resulting shift in affective relations between objects" (Shaw 2012: 614). Thinking in these terms, we might instead ask: if a tree falls in a forest and no human is around to hear it, does it make a difference? It would be fair to say, for example, that if the tree were to land on a squirrel walking about the forest floor that both tree and squirrel might in some way be 'perturbed' by the event of the tree falling (on 'perturbation', see Ash 2013). Equally, the vibrations might reverberate around the forest and while not 'heard' might again have some kind of impact upon the full range of objects present in that situation; sound can be felt in ways that don't involve ears.

The key point is that objects exist and (inter)act whether humans are present or not, perceive it or not, or even exist or not. As Ash (2013: 24) notes:

> a rainstorm can radically perturb a plant in a forest, providing it with the energy for growth, the simple presence of technical objects within an environment can create atmospheres that potentially block, crash, welcome, shortcut or interfere with the circulation of perturbations of objects in all manner of ways.

The key question for OOO then is not what objects means for us, what they do to us, or how they can be used by us, but rather, more broadly, how are objects in the world regardless of us? This isn't to say that they might not mean something for us, do something to us, or be used by us. Obviously, this is the case. The point is that our relations with them do not exhaust their existence or actions; there is far more going on than meets our eyes (or ears).

Unfolding material realms

Thus far, this chapter has been focused on the different debates that have emerged around materiality and NRTs and the different conceptual lenses through which materiality has been (re-)thought. This section will show more how such ideas have translated into geographic considerations of materiality in more worldly situations. This isn't to set up a false distinction between theories and empirical contexts. The ideas discussed here have precisely been orientated toward opening up an expanded material realm of 'real-world' stuff. Rather, the intention is really to show where these ideas have taken geographic investigations of social-material worlds. In doing so, this will look at a range of studies of technology and how, in a range of ways, such technologies unfold through networks of relation between material objects which play roles in processes of materialization.

One of the areas of work where NRTs have been most sensitive to a variety of socio-economic developments has been work interested in various forms of technology, and particularly how such technologies have (increasingly) impacted upon perception and action in social spaces (see Thrift 2004c). Such technologies have been seen to present "a vital component of understanding life itself" (Thrift 2008: 154). Early in the development of NRTs, we find that "Thrift emphasises the role of technology in producing a 'machinic complex' in which an avowedly human experience is displaced by that of the 'cyborg', some kind of hybrid of machine and organism" (Murdock 1997: 731 [citing Thrift]). There is again here an emphasis on relations and a situation whereby any 'organism' cannot easily be detached from its environment; we find ourselves in a situation of "combinatorial production" whereby various technologies come to actively shape the environments in which humans (and others) live, and shape those humans themselves (Thrift 2008: 155). From this, environments become "extended and more active" through the presence of such technologies meaning we are in a period of radical change when it comes to how the world is 'disclosed' (Thrift 2008: 154). Much of this, though, happens without a great deal of thought or reflection. There is, to a degree at least, an increasing ubiquity of such technologies in our day-to-day lives (Kinsley 2012).

Materiality and perception

One way in which such change can be seen is in the role of technologies in shaping human perception and possibilities for action. This can relate, for example, to environmental perception (Wilson 2011a, 2011b) and practices of remembering (Kinsley 2015; Wilson 2015). Ash (2010b: 414) argues that "technology not only shapes the unconscious knowledges humans have about the world", as Thrift argues, "but [also] that technology operates to organise the material sensory capacities of the body itself, and thus the limits of the sensible". Ash (2010b: 415) describes this in terms of 'teleplasticity' which he defines as "molding or forming at a distance". Ash argues that the case of computer games provides a particularly amplified teleplastic context (also see Ash 2015a). This might initially seem counterintuitive given that it is common to read or hear of concerns about the sedentary nature of game play and how this is stopping young people from engaging in more active pastimes. However, this teleplasticity arises given the ways in which the human brain – and so capacities for thought and action – is shaped through relations between bodies, gestures, and the technological interface used, especially where such interfaces require bodies to develop complex motor skills.

To explain this, Ash (2010b) considered two aspects of video game play in the context of the games 'Lego Star Wars: The Original Trilogy' and 'Burnout 3: Takedown': how the limiting framework of the game encourages users to act without thinking and what techniques users develop in response to such limited environments. In terms of the former, video games commonly present a range of limitations on users both in terms of the interface used to play the game (i.e. the controller) and in how actions in the game-world itself are circumscribed. That might relate to anything from the number of buttons on a controller to the speed at which a character can run or the distance that they can jump (see Ash 2010a). Such factors are clearly predetermined by those who design game peripherals and the game itself. Taking the example of 'Lego Star Wars', Ash suggests that such parameters start to matter in terms of the sorts of actions it encourages bodies to engage in. As the avatars controlled by the gamer can 'respawn' (come back to life) an unlimited number of times, this encourages users to keep moving even if they are not entirely sure where they should be going or whether that is the right thing to do. Through this, limitations or boundaries are tried and tested rather than thought through or consciously planned out. Moving to the latter, it is also the case that in playing video games embodied capacities for action take shape in response to those constraints. Taking the example of 'Burnout 3' (a racing game), Ash suggests that gamers adapt their visual focus in light of the speed of the action taking place in the game and to avoid crashing into obstacles or other users. Here, users looked to the middle distance (above the car) rather than at the car they were controlling. Further, when it comes to manipulating the controls, users controlled their vehicles through a series of staccato actions; Ash found that such quick manipulations took place, on average, once every two seconds while playing the game.

What we see here then is how "environments and interfaces can imbue users with a new range of capacities with which to sense space and time, and to orientate themselves" (Ash 2010b: 415). In response to limitations and actions brought about in the interaction with a specific type of technology, through a process of 'somatic reorganization', specific embodied responses and capacities emerged.

Nanoscale materialities

It is not only human perception and possibilities for action that are being shaped through recent technological developments. The material realm itself is being extended as new possible technological developments enter onto the horizon. This can be seen in recent interest in, and concern over, 'nanotechnologies'. As Anderson *et al.* (2007: 139) suggest:

> Posed between dream and reality, and suspended between the future and present, nanotechnology has been heralded as a technology that may define the twenty-first century. Based on claims of either the ability to precisely control and manipulate the material world at the nanoscale, or the ability to modulate living processes at the nanoscale, nanotechnology is said to promise a set of transformative applications that will disrupt established categories such as the artificial/natural or the biological/informational.

The nanoscale refers to a particular unit of measurement and so defines a particular scale of material concern. Research on the nanoscale is interested in the particular properties of the space between 1 and 100 nm. This is a space where all there is is matter, when "What exists, and all that can exist, are atoms and different relations between atoms" (Anderson 2007: 157). Life, then, is reduced to nothing but matter itself, and the differentiation of human and non-human, between identified objects or bodies, becomes unbounded. Here, we have "a multiplication of matter into an excess of new or altered materialities" whereby what matters is not the identity of matter but matter's properties and capacities (Anderson 2007: 157). At this scale, it is argued that "novel properties ... emerge" and so "the novelty of nanotechnology is defined by the uniqueness of a particular space" (Anderson *et al.* 2007: 139). While presented as a horizon of future opportunity or something from science fiction rather than something happening right now – for example, think of Iron Man's suit in *Avengers: Infinity War* which no longer requires the assembly of large panels of mental but rather materializes from a single small chest-mounted apparatus – the ultimate concern is to realize control over the structure of matter itself.

Nanotechnology raises a range of implications when it comes to humans' relationships with the material world and many of the technologies we are currently routinely entangled within. As Anderson *et al.* (2007: 139 [citing UNESCO]) point out, this raises a range of opportunities and concerns, the specific materializations of which we are not yet certain on:

> Nanotechnology *could* become the most influential force to take hold of the technology industry since the rise of the Internet.

Nanotechnology could increase the speed of memory chips, remove pollution particles in water and air and find cancer cells quicker. Nanotechnology could prove beyond our control, and spell the end of our very existence as human beings. Nanotechnology could alleviate world hunger, clean the environment, cure cancer, guarantee biblical life spans or concoct super-weapons of untold horror. Nanotechnology could be the new asbestos.

As such, nanotechnology potentially identifies a host of currently unknown or unnamed materialities which have different properties of size, weight, and strength than those we currently know of. It also identifies materialities with different capacities than those we currently use in sustaining ourselves and our environments. This might lead to new materializations whereby bodies become stronger, healthier, live longer, and are more resilient in the face of threat. It also means that a range of consumer technologies might become faster, more powerful, and so more seamlessly integrated with our environments and with us. And it means that the technologies through which we are defended and destroyed might become lighter, smaller, faster, more independent of human involvement in their operation, and so all the more threatening (Anderson 2007).

Ultimately, developments like nanotechnology pose questions for how materiality is understood. While geography has previously tended toward an understanding of materiality that focuses on the tangible and the solid, the implication of nanotechnology is that "the contingent fabrication of new materialities in nanotechnology necessitates a creative rethinking of the category of matter" such as those we've seen with new materialism, OOO, ANT, and the like (Anderson *et al.* 2007: 141).

'Virtual' matters

There is a risk when thinking of technology that the language used can lead us to think of such things as constituting some sort of immaterial realm (Kinsley 2014). Thinking about technologies through metaphors like 'cyberspace' or 'the virtual', while potentially dated now, tends to conjure images of something that exists separate from our 'real' world. Popular films series like 'The Matrix' have played a part here. As such, Kinsley (2014: 365) argues that we "need to move beyond the frictionless immateriality of 'virtual geographies' towards a greater attention to the material conditions of contemporary digitally inflected spatial formations". Such virtual geographies are in fact constructed by a host of material infrastructures which we are entangled with and are co-constituted by, even if we don't realize it. Any apparent 'immateriality' is very much grounded in a range of technologies, networks, codes, and so on that both exist as a material reality and come to matter given particular capacities to influence the users of those technologies (also see Wilson 2011b). This doesn't necessarily lead us to the most inspiring or easiest of research practice. However, while "it is hard, and sometimes dull, to discuss cables, circuits and sensors", these are, nonetheless, "material expressions of a sociotechnical milieu and, as important constituents of infrastructure ... are composed with code and software, bodies and language in the production of space" (Kinsley 2014: 370).

To illustrate this materiality of what might seem to be immaterial, Kinsley suggests the example of sending a text message. When sending such a message:

> rather than being an 'immaterial' process, there is a significant network of matter and energy upon which this 'virtual' activity is predicated. A device, usually a phone, is used to input the message. The physical functions of that device are contingent upon a wealth of highly processed materials, often with complicated origins, enmeshed into complex chemical arrangements and interdependent components. For example, a capacitative touch screen, made from glass and electrically conductive materials, uses the body's electrical capacitance to sense the point of touch ... This is often processed via one of many Application-Specific Integrated Circuit chips in the device, feeding data to other components that process software, which in turn changes the electrical charge within different areas of the screen (pixels) to display images. To 'send', the software engages the modem of the device to communicate with the network, translating data into modulated pulses in particular frequencies of the electromagnetic spectrum. The phone network infrastructure itself is constructed of many transceivers on towers, providing wireless communications capacity, that are also connected to the physical network infrastructure, which is composed of copper and fibre optic cables, exchange points and network switches, through which binary data are transported over significant distances at the speed of light. Protocols encoded into devices and software control the communication of particular kinds of data. For text messaging, the Short Message Service (SMS) protocol is used by the various actors in the network to route and interpret a message. A centralized SMS Centre, a specific type of server (both hardware and software), stores and forwards messages when the network completes the connection to the recipient device.
>
> (Kinsley 2014: 375)

In the routine act of sending such a message, then, the human body is enrolled into a series of relations with a variety of material systems – both those that it is aware of interacting with and a host of non-human material interactions that take place away from such consciousness (also see Ash 2013) – which come together in recording and conveying meaning both within the system and to another human (the recipient). Or as Kinsley (2014: 375) surmises:

> Circuits of connection, energy transfer, cognition and meaning are composed between bodies, devices, infrastructure, data and others into a milieu that not only performs the action of sending a text message but changes the composition or states of the variously composed individuals involved.

Conclusion

This chapter has introduced NRTs' concerns with materiality and has situated those concerns in relation to a host of recent calls for geography to 'rematerialize' its studies of society and the cultural. In contrast to those calls which tend to equate the material with something solid, foundational, or 'thingly', NRTs have sought to re-think matter in a number of senses. First, there is a shift from a concern for matter as something assumed to be fixed or solid toward a concern for processes of materialization. Echoing NRTs' emphases on process, affect, practice, and events, this means that matter is seen to be perpetually emerging amid the encounters that make up everyday life. Second, this has led to calls to expand how matter is understood. Such materializations take place with the properties of any element or state. Here, we encounter a range of 'elemental geographies' that have looked at phenomena like air and its materializations in political and artistic realms.

A range of conceptual developments have been drawn on in such considerations of matter and materialization. While by no means discrete or simple to synthesize, NRTs have found inspiration in writing on ANT, 'new materialisms', various 'bodily materialisms', and, most recently, OOO. Amongst these conceptual framings, a range of recurring tensions and concerns have become apparent. Perhaps, most prominent here has been the status of the human amid such relationality, non-human agency, and potential 'flattening'. That said, NRTs nonetheless remain concerned with the implications of such agencies for the human and so has arrived at a position whereby a host of relations, affects, interaction, and materialization might be going on which humans do not play a part in and/or are consciously aware of, but that is not to say that such interactions and intra-actions do not matter for said humans. In effect, we find ourselves in infinitely more complicated and challenging material situations.

Further reading

Coole and Frost (2010) introduce some of the main features of 'new materialisms' and give an overview of some of the political importance of such conceptual approaches. This covers a wide range of situations, including climate change and biotechnological developments such a GM foods. In that, the division of the natural and the social is called into question and questions of ownership and responsibility complicated. **Robbins and Marks (2009)** provide a critical introduction to various recent calls in geography to 'rematerialise' geography and identifies an 'assemblage geography' in response. Robbins and Marks discuss the limits of recent materialist work, focusing on issues of clarity and expression, writing such relational geographies, and their position relative to other established approaches to thinking about nature–society relations. **Kinsley (2014)** provides an overview of a host of work on technology in geography and highlights how this work has variously understood that in terms of materialities and immaterialies. Kinsley argues both for a recognition of the materiality that lies behind apparently immaterial phenomena (like 'cyberspace') and for a particular take on materiality that looks at the co-implications of technology and humans (discussed in terms of 'technics'). **Shaw (2012)** both provides a useful review of a range of approaches to materiality in geography and develops an account of geography's recent interest in 'objects' and what he calls 'geo-events'.

5

NON-REPRESENTATIONAL
THEORIES AND LANDSCAPE

Introduction

Landscape appears to be a fairly straightforward object of concern. It is commonly defined as a portion of land or scenery which the eye can view at once, or equally as land, or a picture/painting of land, that can be seen from a particular point of view (Duncan 2000). However, landscape has been a much debated area of interest for cultural geography and a key ground upon which a host of ideas around NRTs were initially explored (see Wylie 2007a). NRTs were initially articulated as a response to a particular type of 'representationalism' whereby representations took precedence over lived experience (see Thrift 1996 and Chapter 1). More specifically in terms of landscape, NRTs were presented in contrast to New Cultural Geography's focus on interpreting a host of representations of landscape or looking at landscapes in terms of iconography and text (see Daniels and Cosgrove 1989; Duncan 1990). Crudely, NRTs' approaches to landscape were positioned as a shift in emphasis from representations to practice and embodied experience. This, in turn, has proved to be a contentious line of argument and has led to a lot of debate on how landscape should be studied by geographers.

This chapter will explore the emergence of non-representational approaches to understanding landscape. In doing so, a brief history of landscape research in geography will be provided in situating the differences NRTs sought to make. From there, the discussion will move onto consider the work of Tim Ingold which has been very influential in introducing themes of practice and experience into geographic scholarship on landscape. Attention will be drawn to how, arguably, the term landscape itself implies a fixity and finished character and therefore requires re-consideration. In response, the chapter will look to the questions asked by NRTs about how landscapes come to be animated by people's practical engagement in and with them. Having set that scene, three substantive themes – dwelling, mobility, and haunting/spirituality – will be explored, drawing on a range of scholarship associated with NRTs and the debate that has emerged around them.

Landscape and cultural geography: a brief sketch

To understand the specific nature of the shift that NRTs enacted when it comes to understanding landscape, it is important to have a sense of what cultural geographers have done previously in the study of landscape. Landscape has been a key object of concern for cultural geographers for more than 100 years. In this time, the ways in which geographers have understood and engaged with landscapes has changed quite dramatically (see Wylie 2007a). This section will pick out two key approaches from this history that are useful in understanding the specific focus of NRTs: those of the Berkeley School of Cultural Geography and New Cultural Geography.

Carl Sauer's (2008) concern with the 'morphology' of the landscape presented a significant moment in the history of geography's concerns with landscape. Sauer and a range of followers explored how "human activity was the single greatest agent of landscape change and that land use varies according to cultural preferences" (Kenzer 1985: 267). The focus, then, was on "material manifestations of culture – land use, settlement patterns, technology, and other artifacts" (Solot 1986: 508). In sum, the task for cultural geography under this approach was

> to describe the morphology of the landscape – that is, the shape, form and structure – of a given landscape, and in so doing to reveal the characteristics, trace, distribution and effectivity of the human cultures that had inhabited and moulded it.
>
> (Wylie 2007a: 23)

Sauer and this Berkeley School of Cultural Geography held a prominent place in US-based cultural geography for a significant part of the 20th century. However, as noted in Chapter 1, a range of critiques of the Berkeley School emerged by the 1970s given a range of social processes and tensions coming to a head at that time. As Don Mitchell (2000: 35) noted:

> As America (and British) cities burned in the wake of race riots, the collapse of the manufacturing economy, and fiscal crisis upon fiscal crisis, American cultural geographers were content to fiddle with the geography of fenceposts and log cabins....

Such concerns led to the development of a quite different type of landscape geography which came to be referred to as New Cultural Geography. New Cultural Geography was pursued by a range of geographers drawing on different ideas which evolved over the course of the final decades of the 20th century and the start of the 21st century. Sources of inspiration included the emerging Cultural Studies of the time (Jackson 1989), Cultural Marxism (Cosgrove 1984; Daniels 1989; Mitchell 1996), and both structuralist and post-structuralist philosophies (Barnes and Duncan 1992; Duncan and Duncan 1988; Matless 1998), as well as feminist scholarship (Rose 1993). Common to this work was a concern for how culture is spatially constituted and so a concern with how issues of power, ideology, discourse, and forms of representation were enacted within space. Landscapes were variously seen to be duplicitous, meaning-filled, iconographic, and open to multiple interpretations.

NRTs approach landscapes differently from each of these schools of cultural geography. While the Berkeley School is evidently interested in materiality and the human activities which come to shape the morphology of the landscape, there is little focus on, or sense of, that shaping actually taking place through specific human action or of those materialities playing much of a part in that shaping. If for Sauer (2008: 103) "Culture is the agent, the natural area the medium, the cultural landscape the result", the unfolding of 'culture' here is vague; a black box rather than a realm of practical action that geographers might explore. Further, when it comes to NRTs' differences from New Cultural Geography, NRTs question the largely visual orientation of such landscape studies. For New Cultural Geography, landscape was "quintessentially *visual*" (Wylie 2007a: 55). There is diversity in that. It might mean landscapes being approached in terms of what is seen by a laborer or in terms of how an artist represents a landscape. It might also be related to whose realities are represented or not in such images/ imaginations. The common theme here, though, is around visuality and how landscapes are imagined and represented (Wylie 2003). By way of contrast, NRTs seek to "think of landscape in verbal terms ('to landscape')" and so "involves phrasing landscape in terms of action, activity and performance: landscape is a doing, to landscape is so to *do* something" (Wylie 2007a: 121). Vision remains a part of NRTs' concerns (see Wylie 2006b and below). The point here, though, is to view "Landscape as 'lifeworld', as a world to live in, not a scene to view" (Wylie 2007a: 149). Thinking in terms of lifeworld draws attention to our 'taken for granted' habitual actions that unfold in our practical, embodied experiences (Ley 2009; see Chapter 1). In sum, for NRTs:

> landscape is conceptualized in terms of active, embodied and dynamic relations between people and land, between culture and nature more generally. The general argument is that landscape comprises the totality of relations between people and land. These relations are seen as ongoing and evolving rather than static, they constitute an embedded and engaged being-in-the-world that

comes before any thought of the world or of landscape as merely an external object. Body and environment fold into and co-construct each other through a series of practices and relations.

(Wylie 2007a: 143–144).

The remainder of this chapter, then, will explore various senses of this folded, unfolding co-constitutedness between body and landscape.

Temporality and animating landscapes

A key tension that NRTs' approaches to landscape make clear lies between certain visions of or for landscape which take landscape as something that is already accomplished (and therefore available for the researcher to reflectively deconstruct) and a version of landscape which is more open or dynamic. This tension isn't just confined to work inspired by NRTs. For example, Tim Cresswell (2003) has questioned the temporality of landscapes in the context of a concern for mobility. As Cresswell (2003: 269) suggests:

> I do not use 'landscape' as a part of my regular academic vocabu-lary. Frankly, I am not keen on the term... Landscape ... does not have much space for temporality, for movement and flux and mun-dane practice. It is too much about the already accomplished and not enough about the processes of everyday life.

Cresswell proposes the term 'landscapes of practice' in an effort to intro-duce a greater sense of movement into the picture. While such a construc-tion might appear "somewhat oxymoronic", Cresswell (2003: 270) argues that it introduces a sense of "negotiation between continuity and change" as "Landscape ... appears to encapsulate the notion of fixity – of a text already written – of the production of meaning and the creation of dominating power" and "Practice ... is about fluidity, flow and repetition". The emphasis is on how landscapes are practiced and animated through such practicing (see Merriman et al. 2008).

Thinking in slightly different terms, Rose and Wylie (2006) approach the issue of the fixity of landscape by focusing on a series of tensions which might be seen to animate it. Writing in the wake of a range of nonrep-resentational work which has emphasized performance, process, materiality, embodiment, and so on, Rose and Wylie express concern over a series of directions here which might be seen to trouble landscape. The 'topologi-cal' metaphors of networks and assemblages discussed in Chapter 4 tend to focus on the connections that compose those networks rather than matters of distance or position. This, in turn, has a tendency toward 'flattening' the world in terms of assumptions about agency or the prior status of that which enters into a relation (see Chapters 2 and 4). This, Rose and Wylie suggest, presents at least three concerns when it comes to landscape, specifically in terms of the distance traveled by such theories from how landscapes are often understood, namely that (i) landscapes, and the language used in discussing them, are framed in static terms rather than being composed in and through relations; (ii) landscape is often associated with a particular version of the cultural construction of nature which has been succeeded by recent forms of 'biogeography' which deconstruct such nature–society binaries; and (iii) landscapes are often understood through topographical rather than topological terms, through measurement, distance, surfaces, relief, and so on rather than horizontal connections. Such tensions mean that "it seems difficult to accommodate landscape, with all its topographical, visual, phenomenal, presentist, and synthetic associations, within a topological vision" (Rose and Wylie 2006: 476).

Rose and Wylie suggest that matters can be looked at in reverse, also. More specifically, they argue that:

> in prioritising vectors, trajectories, and connections, topological and vitalist geographies present a curiously flat and depthless picture – and it is here that notions of landscape, perception, and subjectivity potentially reemerge.
>
> (Rose and Wylie 2006: 476)

Against such 'overflattening', Rose and Wylie (2006: 477) suggest that landscape might be seen to reintroduce "perspective and contour; texture and feeling; perception and imagination" in that landscape "is in the synthesis of elements, so elegantly traced by topographies, with something added: lightless chasms, passing clouds, airless summits, sweeping sands". And in pursuing this, Rose and Wylie suggest the pursuit of what they call a 'post-phenomenological' conception of landscape (also see Rose 2006; Wylie 2006b; Box 5.1). Such a position here does not enact the same flattening as the relational approaches that focus on networks and assemblages. At the same time, though, things like 'the subject' that might survey the landscape they occupy are, and unfold through, their embodied relations with the "textured topographical presence" and horizon that is landscape (Rose and Wylie 2006: 479). It is not so much the case here that landscapes need to be animated and rather that landscapes animate cultures, bodies, subjectivities, amongst other things. Or to put it in other terms, here, "landscape names the creative tensions of selves and worlds. This is the animating quality of landscape" (Rose and Wylie 2006: 479).

BOX 5.1 'POST-PHENOMENOLOGICAL' GEOGRAPHIES?

The challengingly named 'post-phenomenological geographies' refers to a set of ideas that have emerged from work generally associated with NRTs and initially, if not exclusively, developed from work on land-scape and embodied encounters in/with landscapes (see Lea 2009). This work, more clearly than other areas of NRTs, shows both the importance of humanistic geography to NRTs and at the same time NRTs' departure from such work (Ash and Simpson 2016; see Chapter 1). Part of the inheritance comes in its return to phenomenological ideas and the use of these in thinking about the embodied experience of the world (also see Simonsen 2005, 2013). However, there is a departure here from phenomenological/humanistic ideas in at least two senses. First, in these post-phenomenologies, there is "a move away from the assumption of a subject that exists prior to experience towards an examination of how the subject comes to be in or through experience" (Ash and Simpson 2016: 49). Post-phenomenological geographies do not take a subject-centered approach to thinking about the experience of space and place. It is less the case for post-phenomenological geographies that there is a subject that makes the world meaningful, and rather, there is a circumstantial coming together of a host of human and nonhuman entities and processes as part of an ongoing 'worlding' (see McCormack 2014, 2017). Second, as part of this less subject-centered approach, there is, then, "a recognition that objects have an autonomous existence outside of the ways they appear to or are used by human beings" (Ash and Simpson 2016: 49). This means that much more agency is given to the world of stuff, and it is not the case that a human subject need perceive or be aware of that agency for it to exist or matter (see Chapter 4; Roberts 2019a). Rather than a complete flattening out of the world into a set of topological relations as has been the case in a range of post-human and relational geographies, in such post-phenomenological geographies, there is both a retention of a concern for subjectifications and a recognition that, as such, subjects are not central to the unfolding of the world (Ash and Simpson 2019).

A key source of inspiration for such work on the co-constitutedness of bodies and landscapes – the parallel emergence of person and world – which in turn emphasizes the dynamism of this relation over a static image of a landscape has been the writings of the anthropologist Tim Ingold (see Ingold 2000, 2011, 2015). Ingold, in fact, articulated his take on landscape in direct contrast to those of New Cultural Geographers. For example, Ingold disagrees with Cosgrove and Daniels's (1988:1) claim that landscapes present us with "a cultural image, a pictorial way of representing or symbolising surroundings". In response, Ingold (2000: 191) rejects:

> the division between inner and outer words – respectively of mind and matter, meaning and substance – upon which such distinction rests. The landscape, I hold, is not a picture in the imagination, surveyed by the mind's eye, nor however is it an alien and formless substrate awaiting the imposition of human order.

We return here to a critical relationship with the sort of representationalism that NRTs were initially positioned against. There is a desire to move away from any notion of a body as separate from the world either in terms of mental images or in terms of established external realities. In response, Ingold (2000: 173) argues that "It is through being inhabited that the world becomes a meaningful environment". Both entities – landscape and person – perpetually emerge and take shape in the unfolding of life. Landscape is "with us, not against us" and through our living in this way, we are part of it (Ingold 2000: 191). In exploring such inhabitation, Ingold turns to a range of sources, including the radical re-thinking of ecological ideas of inhabitation found in the writing of James Gibson and the existential phenomenology of Martin Heidegger (also see Ingold 2011). This is all articulated in terms of how the landscape is dwelt within.

Dwelling in the landscape

One of the key lenses through which inhabitation and co-emergence have been explored is that of dwelling and the dwelling perspective (see Jones 2009). Cloke and Jones (2001: 651) summarize such an approach as follows:

> Dwelling is about the rich intimate ongoing togetherness of beings and things which make up landscapes and places, and which bind together nature and culture over time. It thus offers conceptual characteristics which blur the nature-culture divide, emphasize the temporal nature of landscape....

This emphasis on "how humans and other animals make and inhabit 'lifeworlds' through registers of specific bodily practice" has meant that dwelling has formed a key concern for geographers developing non-representational approaches to landscape (Jones 2009: 266). That said, this is not geography's first engagement with ideas around dwelling. Humanistic geographers also drew on ideas around dwelling to think through people's belonging and togetherness in terms of "underlying patterns, structures and relationships" (Seamon 1993: 16; see Chapter 1). However, work around NRTs has seen in ideas of dwelling the potential to develop an account of human–environment relations which is "radically relational" rather than based on underlying structures and, in this, is more anti-, post-, or trans-humanist in orientation (Harrison 2007:626). This interest in practices of inhabitation has entailed engagements with various conceptualizations of dwelling as well as a range of critical reflections upon such understandings. In exploring such work, this section will begin with the distinction drawn between 'building' and 'dwelling' before moving onto some more critical engagements with these ideas.

Building versus dwelling

One of the ways that Ingold (2000) explains the particularities of dwelling is through a contrast to what he calls a 'building perspective'. Such a building perspective approaches the landscape in a similar way to how Berkeley School Cultural Geography understood the actions of culture on the landscape: "Reality, that which is imposed upon, is envisioned here as an external world of nature, a source of raw materials and sensations for diverse projects of cultural construction" (Ingold 2000: 178). According to this perspective, the world is made or constructed *before* it is lived in. Such construction is planned, and such planning precedes its actual expression in material form (shelters, services, roads which connect them, the overall organization of all of this, etc.). There is a separation of human thought and action from the world (Jones 2009), a separation between a perceiving subject (the human) and the environment to be engaged with or made home. Ultimately, this renders that which is built as little more than a container for the activities of living.

By way of contrast, from a dwelling perspective, the world is made through the acts of living in it and, as such, is never complete (Ingold 2000). As Jones (2009: 267) explains, here, "any form of life emerges from the world and there is never a gap, or break, in which thought and practice can completely free itself". We lose any sense of dualism between thought and world and between form and process. In developing this point, Ingold (2000: 186 [emphasis in original]) draws on the work of the philosopher Martin Heidegger who argued that:

> We do not dwell because we have built, but we build and have built because we dwell, that is because we are dwellers ... To build is in itself already to dwell ... *Only if we are capable of dwelling, only then can we build.*

Figure 5.1 The Harvesters (From https://www.metmuseum.org/art/collection/search/435809).

The key point here is that building arises from within the unfolding of activity and a specific engagement with the surroundings of that activity. Any advanced planning or imagining here is not detached from the world but rather is undertaken by people who are *already in the world*. Ideas are not 'imported' from the outside but rather emerged from within the world. These ideas are only possible because we have already dwelt (Ingold 2000). Such an emphasis on the co-constituted nature of human–landscape relations here again draws greater attention to the agency that exists on the side of the landscape and the various things that exist within it. 'Nature' itself is an active unfolding field, and a range of things exist within it that form relations and afford various sorts of activities in the unfolding of dwelling. This is lost in the building perspective where 'nature' is something to be conquered and mastered rather than lived with.

To illustrate the dwelling perspective, Ingold analyzes Brugel's painting 'The Harvesters' (see Figure 5.1). By looking at this painting, Ingold seeks to illustrate the ways in which a landscape is much more than a mere (symbolically ordered) backdrop against which human activities unfold. Instead, Ingold seeks to show how the landscape is an enduring record of the lives of people who have dwelt therein. In approaching this painting in this way, Ingold (2000: 202) asks his readers to approach a reflection on the painting in a very particular way:

Rather than treating the world as its own painting I should like you, reader, to regard this painting by Bruegel as though it were its own world, into which you have been magically transported.

Image yourself, then, set down in the very landscape depicted, on a sultry August day in 1565. Standing a little way off to the right of the group beneath the tree, you are a witness to the scene unfolding about you. And of course you hear it too, for the scene does not unfold in silence. So used are we to thinking of the landscape as a picture that we can look at, like a plate in a book or an image on a screen, that it is perhaps necessary to remind you that exchanging the painting for 'real life' is not simply a matter of increasing the scale. What is involved is a fundamental difference of orientation.

Having undertaken this shift, Ingold draws attention to a range of features in this scene which show the co-emergence of people and landscape and the dynamism that is present even in the most apparently secure or atemporal elements present. For example:

- While the hills and valleys suggest a sense of permanence, they are, in fact, taking place on inhuman timescales: "As you watch, the stream flows, folk are at work, a landscape is being formed, and time passes" (Ingold 2000: 203). Further, our understanding of these features comes less at the level of abstract measurement or classification and more through our felt engagement with them; the physical act of ascending and descending gives us our understanding of their contours as does the movement of our vision across them.
- The paths and tracks show how our arrival into a landscape and a specific point of view on/within it come from a range of such routine movements through the landscape that people undertake. There is again a co-constitution here in that such routes emerge from the "accumulated imprint of countless journeys that people have made...as they have gone about their daily business" (Ingold 2000: 204).
- The pear tree marks a key place in the scene in that:

the tree bridges the gap between the apparently fixed and invariant forms of the landscape and the mobile and transient forms of animal [i.e. human] life, visible proof that all of these forms, from the most permanent to the most ephemeral, are dynamically linked under transformation within the movement of becoming of the world.

(Ingold 2000: 204–205)

The tree shows a history unfolding in relation to its context from the point at which it took root; it has been nurtured, pruned, picked, weathered, leant against, and so on.

- The corn fields draw attention to the productive relationships between people and land that are so often fundamental to human inhabitation. In this, we can get a sense of the rhythms of dwelling – of planting, tending, harvesting, refining, consuming, and so on.
- The church in the scene might be less 'natural' than the tree and was obviously built; but here, Ingold sees its ongoing presence to be intertwined with practices of dwelling in the landscape. As the tree has roots buried in the soil, so do the people who bury their dead in the graveyard. The church also comes to be entwined with non-human processes in the landscape in terms of it being weathered by the elements and inhabited by animals and plants.
- The people in the scene make sounds through their activities – harvesting, eating, snoring, talking, the church bells, and so on – that fill and animate the landscape. These people are variously engaged in practices of landscaping, both literally shaping the landscape but also making it meaningful through shared practices (though see Box 5.2).

BOX 5.2 QUESTIONING DWELLING AND SLEEPING HARVESTERS

While ideas around dwelling have been influential in how geographers have developed thinking about human–landscape relations, critiques of this work have emerged that echo of the sorts of discussions encountered in Chapter 2 around NRTs' (overly) lively account of practices. In particular, Harrison (2009) re-reads 'The Harvesters' and questions Ingold's focus on the activity of practice by focusing on the sleeping figure in the painting. Harrison develops this re-reading in the interest of both addressing a substantive gap in NRTs' discussions of practice (and landscape) – sleep is a subject that has received limit attention – but also as such a reading potentially sheds light on certain analytic blind spots that a focus on and/or valuing of activity brings. As such, "rather than considering the practical disclosing or becoming of worlds", Harrison's (2009: 989) account "may be understood as a reflection on what happens in the suspension of all comportment, disposition, and posture".

The sleeping figure in 'The Harvesters' is almost completely absent from Ingold's (2000) account of the painting and the practices of landscaping he sees within it. The sleeping figure is only mentioned twice. The first mention comes as part of a broader discussion of the sounds of activity (here snoring), and so the sleeping figure is not specifically mentioned for its status as a sleeping body. The second mention argues

(Continued)

---

that the sleeping figure is 'out of joint' or does not fit with the social rhythms of activity that pervade the scene and through which the world becomes. For Harrison, that relative lack and limited consideration is not a simple oversight. Rather, it is an effect of the specific analytic frame being developed. Ingold's "prioritisation of practice necessarily renders insignificant or disavows the susceptibility and finitude of corporeal existence, sublating these dimensions of corporeal existence into a dynamic-relation account" (Harrison 2009: 991). In such an analysis, being 'out of joint' is not something understood or recognized for its significance in understanding our embodied situation and rather is something that is seen to be secondary to being 'in joint'.

In response, Harrison seeks to re-think how we see and value susceptibility and/or finitude such that they are not seen as deficiencies in practice. Rather, such states are the conditions of possibility from which activity emerges. As Harrison (2009: 995) notes, "The active subject, the subject of practice, is possible as such only as a subject who sleeps, who has slept, and who will sleep again". This presents us with quite a different valuation of what is going on in the scene and a recognition that such a sleeping figure is not, in fact, 'out of joint' with what is going on. Rather, sleep is common to all present in the scene itself. Further, in this status of sleeping, we find a withdrawal or separation from what is going on around the figure but not in a way that positions this as a negative or lack. Rather, sleeping is a point from which inhabitation unfolds; again, separation is not something negative in terms of being detached from a collective or community of practice, but the point from which such a collective is approached. Harrison terms this as 'intervallic sociality' which is significantly different to what we might commonly think of when it comes to notions of community as a cohesive thing.

Moving on with dwelling

While dwelling has proved to be a key source of inspiration for geographers who are interested in practice and landscape, a range of critical considerations have emerged in response that push discussions of landscape in a range of directions (see Rose 2012). For example, Cloke and Jones (2001) question the common distinction found in work on dwelling between 'authentic' and 'inauthentic' forms of dwelling. This distinction needs to be addressed as it presents us with "a sinister (nationalistic) rustic romanticism" (Cloke and Jones 2001: 611). Here, there is commonly a value-laden distinction made between authentic and inauthentic modes of dwelling, with the former being seen to be 'proper' and the latter deficient in terms of the engagement between people and landscape. This, in many ways, mirrors some of the themes in the building versus dwelling discussion. Such a distinction becomes problematic also when we start to think in terms of our current societal context which is, in many cases, quite removed from the sort of pastoral–agricultural settings that Heidegger and others were referring to when first discussing these ideas. As Harvey (1996: 301) questions, "what might the conditions of 'dwelling' be in a highly industrialised, modernist, and capitalist world?" By way of response, Cloke and Jones (2001) develop a revised understanding of dwelling by drawing on a range of relational ideas, including ANT (see Chapters 1 and 4). Ingold's accounts of dwelling do in themselves emphasize relationality in their emphasis on contextualized involved activity (see Ingold 2011). However, Cloke and Jones (2001) take this further in focusing on the hybrid assemblage of different tasks that unfold in the inhabitation of a dynamic/unfolding 'taskscape'. In their case, this 'taskscape' is a Somerset orchard where 'traditional' and 'modern' practices of cultivation co-exist and overlap. This illustrates the problem of trying to talk in terms of 'authenticity' in that even the 'traditional' practices at some point were 'modern' innovations. In this sense, in their orchard:

> we do not uncover a sterilised museum of past landscape and dwelling, somehow untouched by current technologies and practices. Instead, we see a series of practices which have evolved over time, and changes which are constantly informed by shifting economic, technical, and cultural formations; we see a place that is not conductive to fixed-point notions of authenticity.
>
> (Cloke and Jones 2001: 657)

Further, Wylie (2003) asks questions about the sensory registers that dwelling is (and isn't) thought through, identifying a push back against visually orientated understandings of landscape. Rather than moving away from a focus on vision, Wylie suggests that vision should in itself be understood as an embodied practice. In this, Wylie (2003) identifies a problematic binary between visual accounts of landscape – generally taken to refer to those accounts focused on representations of landscape and more broadly the 'representationalism' NRTs find problematic – and those identifying embodiment as a corrective focus which is variously multi-sensory (but generally not visual) in orientation. Wylie suggests that a more thorough re-thinking of vision and seeing needs to be undertaken that recognizes that the act of looking is itself part of dwelling. Seeing need not, for example, be a matter of detached oversight undertaken by an individual that is in some way removed from the landscape being looked at. Rather, Wylie (2006b) suggests a shift in orientation toward thinking about how we see 'with' the landscape and the depth it presents us with, and that this in turn presents a creative tension whereby self and world are folded/unfolding (see Rose and Wylie 2006). This can be seen in Wylie's account of the feeling of vertigo he experienced at the summit of Glastonbury Tor in South West England. Here, there is a 'vertigo of disconnection':

> On the summit of the Tor, one is carried 'above', one is no longer 'on the ground'. And yet the summit is revealed as a relatively broad platform, whose borders are rounded rather than sheer. One could never 'fall off' the top of the Tor. The true reason for this feeling of vertigo is the visible panorama itself. The field of vision is unaccountably broader and deeper than the rigour of the ascent would lead one to expect. All its dimensions seem exaggerated. The sensation of height is uncanny: more of the visible than there should be. In every direction the recumbent fields of the levels advance into depth. No clear-edged horizon arrests this onward and upward advance, and those who stand upon the Tor see depth in its clarity and inexhaustibility. On the summit the visible world assaults vision. One sees to the limits of seeing, and this is vertiginous. It is not the vertigo of losing touch with the earth, it is the vertigo of being displaced from 'oneself', from a habit of considering one's sight to be the vehicle of an interiorised intention. To see to the limits of one's vision – to be encompassed within visible depth – is to recognise that sight is etched out of landscape, rather than the landscape being the product of sight. This is not just an acknowledgement of the landscape's material presence, or its persistence beyond the horizon. To perceive is to be already and always caught by dwelling.
> (Wylie 2003: 153)

Landscape here is not what is seen but rather "names the materialities and sensibilities with and according to which we see" (Wylie 2006b: 520). This, in turn, contributes to the broader push that we have seen throughout the preceding chapters whereby, for NRTs, the subject – here one looking at a landscape – is significantly decentered, something that is not quite so evident in Ingold's account of dwelling (see Wylie 2006b). Rather than being that which enters into and/or orientates relations, subjectivities (plural) become outcomes of relations and processes – here, processes of depth and folding between body and landscape (see Simpson 2017d).

Wylie (2003) also poses questions about the temporalities implicated in Ingold's post-Heideggerian account of dwelling. Again, concerned with the sort of sinister romanticism that can emerge from a focus on sustained engagement and, even, from the ideas of rhythms and cycles of (re)engagement found in Ingold's account of the Harvester (seasons, harvesting, planting, tending, etc.), Wylie (2003) suggests that we might think of dwelling in more transient or fleeting senses as something that might unfold, for example, in the context of walking up a hill like Glastonbury Tor. The question becomes less about the long-term duration of a given human–landscape interaction and instead becomes about dwelling as the *ongoing* practice through which human–landscape interactions are enacted and enabled.

Finally, Harrison (2007) returns to the question of authenticity and inauthenticity by considering a different sort of ethics that might emerge in thinking in terms of dwelling. Harrison draws attention here to both the implications of thinking about dwelling in light of the work of Martin Heidegger – influential in Ingold's account – and to how a different understanding might be arrived at if we look to other sources. In this, Harrison draws particular attention to the centrality of 'being-at-home-in-the-world' to such Heideggerian accounts and suggests that it might also be possible to think about dwelling in terms of an exposure to otherness. In Heidegger's writings on dwelling, Harrison identifies a range of tensions and impulses which relate to how individuals might be 'abroad in the world' but, in this, seek enclosure and boundaries (also see Rose 2006). Looking to the writings of Emanuel Levinas, Harrison introduces ideas of alterity and spacing – as opposed to belonging and connection – and how this can be seen to be a constitutive (rather than privative) relation to dwelling. Such an emphasis on alterity gets us away from the emphasis on 'proper' modes of dwelling or the sort of sinister romanticism Cloke and Jones (2001) signal. That is, such an emphasis on alterity does not deny that these romantic ideas and aspirations might be pursued, but rather that they miss something in our relations with the world and others therein by covering over a prior circumstance of being in relation with others. On that basis, rather than nostalgic desires for authenticity or rootedness, we are presented here with an "unsettling proximity" which is not something to be resisted but rather is an encounter that requires response (Harrison 2007: 635). Here, an encounter with an other calls into question any 'being-at-home-in-the-world' and such openness, such being called into question, represents the orientation of our relationships with our situation. Dwelling unfolds through being troubled in relation.

Walking through the landscape

Returning to the issues raised by Wylie (2003) in terms of fleeting and dynamic forms of relation between people and landscapes (which may or may not be discussed in terms of dwelling), a key practice through which such human–landscape relations has been considered is walking. While walking has received quite sustained academic attention in recent times (for example, see Edensor 2000; Pinder 2001; Solnit 2001), walking has offered those developing NRTs a specific context in which to consider embodied practices of being in the world. In this section, walking will be considered in three senses. First, walking will be looked at as an embodied practice that suggests various registers of experience unfolding between body and landscape. Second, the archiving and representation of such walking practice will be considered. Finally, the issue of bodily difference will be broached in an effort to recognize the importance of the different capacities bodies have to affect and be affect in and by landscape.

Walking in the countryside

Wylie's (2002a, 2005, 2009) accounts of walking through landscapes have been significant in the development of non-representational accounts of body–landscape relations. In these accounts, we encounter experimental attempts at presenting the different sensibilities, movements, affects, feelings, and moods that emerged the context of various walks taking place in the South-West of England, often on the 'South-West Coast Path'. The South-West Coast Path runs for 630 miles around the coastlines of the counties of Somerset, Devon, and Cornwall. Wylie's agenda in walking the Path was not to study its social–historical geography – not to broach, for example, matters of access/exclusion or appropriate conduct, of who walks and why (though see Box 5.3) – but rather to walk in an effort to "activate a space and time within which [he] might engage with and explore issues of landscape, subjectivity and corporeality" (Wylie 2005: 234).

BOX 5.3 ALTERNATIVE TAKES ON WALKING THE SOUTH-WEST COAST PATH

Wylie's (2005) account of a single day's walking on the South-West Coast Path generated a range of critical responses amongst geographers interested in landscape. For example, in direct response to Wylie's (2005) account, Mark Blacksell (2005: 518) has suggested that he found Wylie's analysis of "what shapes our reactions to landscape quite fascinating for the way in which it illuminated the subtleties behind our individual perceptions" and, as such, how "Who we are, and where we come from, are … key determinants of what we perceive and how we react to any situation". This sort of focus, for Blacksell (2005: 518),

---

holds the potential to act as "a powerful antidote to the didactic world of the guidebook, with its aim of imposing a pre-set structure on what we see, smell and hear". However, Blacksell (2005) suggests that such an approach risks a 'narrowness' in its introspective mode of analysis and calls for a greater concern for a variety of contextual concerns which might shape such an encounter with a landscape: literary and artistic representations of such or similar landscapes; guidebooks and various topographical descriptions of the landscapes moved through; the practices of landscape management and preservations that play a part in shaping what is encountered; and, the cultures of working (risk assessment and the like) that increasingly form a pre-requisite before any actual engagement beyond the bounds of a University office. There is, in broad terms, a desire to frame the practice of walking with a range of discourse around, and representations of, landscape.

These sorts of critical comments are amplified in another response to Wylie's paper, this time from James Sidaway (Sidaway 2009). Here, Sidaway (2009) undertakes his own walk on the South-West Coast Path but focuses on a different segment of the Path than runs through Plymouth's urban coastline. In this, Sidaway identifies what he calls 'master keys' which explain the shaping of the topography of this section of the Path and so add to the various contextual-discursive materials suggested by Blacksell. Sidaway discusses Plymouth's military–naval history and present, wartime bombing, and associated questions of (in)security (as well as a series of other matters around urban development/regeneration) which fold with other matters of memory and loss that inflect his attention "to social relations and mediations of property and access, industrialisation, deindustrialization, consumption, gentrification, and geopolitics" (Sidaway 2009: 1107).

Such critiques and extensions raise important points when it comes to the way that work informed by NRTs has approached landscape. It is evident that the question of context is important when it comes to embodied encounters in and with landscapes. Bodies and landscapes have histories which have a bearing on the unfolding of these interactions. However, there is a risk in such a quick push back to representational-discursive contexts that we lose sight of the ground gained and the fact that such work informed by NRTs is precisely trying to rethink how geographers approach landscape. In turn, that suggests that how we think about such contexts and situations needs to be rethought. It is not the case that such contextual matters – in whatever form – foreclose or explain an encounter with landscape or are a

(Continued)

prerequisite to a proper academic (or broader) encounter with landscape. Such a status would significantly denigrate the sorts of embodied encounters Wylie and others speak of, reducing them to some sort of epiphenomenon of a more fundamental interpretive, (pre-)scripted sort of practice or a determinative context that governs experience. Wylie (2005: 521) himself pushes toward such a rethinking in response to Blacksell in suggesting alternative contextual materials that might matter to his walk but which supplement and make strange "the variable, sometimes numbing inheritance of landscape histories and idioms". Here, context is not taken to present "a setting or a filter, or a legacy to genuflect before, but rather something to edge around and sense obliquely, taking care not to let it take over" (Wylie 2005: 521). Or to put this in another way, such contextual matters might be taken in what McCormack (2014) calls a more 'circumstantial' sense rather than as something determining or foreclosing outcomes. Here, what might be taken to be contextual matters become enrolled as part of a broader 'loose gathering' of a whole host of human and non-human 'stuff' that bring about 'fluctuations and deviations' in both self and landscape as something like a walk unfolds.

Such an agenda raises a range of significant points in thinking about walking as an embodied encounter with landscape composed of "moments, movements and events" (Wylie 2005: 236). First, and most broadly, we gain a sense of the unfolding of bodily feelings as they emerge within the context of specific settings or locales and the senses of self and landscape that emerge from (and are disassembled through) them. For example, when it came to walking through a wooded area of the Path, Wylie (2005: 238) recounts how:

> the walker in the woods is straightaway, nervously and anxiously, an encompassed self. In temperate latitudes at least, woodland commonly has a particular density, one which admits enough light to make it clear that beyond this tree, these branches, a tangle of wood and leaf extends in all directions.

Equally, this changes when:

> The woods cleared and for a time the Path found open air and a level course upon a billowing landscape, with fields of pasture running right up to the cliff-edge and the grey-blue sea sponging in the

background. Odd to be edging microscopically along a flat wedge of land that dropped so suddenly and vertically. Then it crinkled into a series of abrupt, densely vegetated rises and plunges, twisting and creasing through steep coombes. I found myself, as happened every day, in the thick of it. In the thick of it: wet, livid green ferns all around, the Path a thin, muddy rope. Limbs and lungs working hard in a haptic, step-by-step engagement with nature-matter. Landscape becoming foothold.

(Wylie 2005: 239)

These variations in the experience of the landscape draw attention to the unfolding nature of bodily feelings that emerge from relations with the surrounding environment. Again, there is a perception with the landscape rather than a distanced perception of the landscape. This is capture by what Wylie's (2005: 242) calls of the unfolding 'levels of sense' through which landscapes are experienced. Such levels refer to a tuning of perception in response to a circumstance rather than a gaze that would perceive some objective reality of that landscape. In this, for example, degrees of light and shade are sensed in the context of a particular circumstance – the unfolding engagement with landscape. The landscape appears (rather than 'is') bleached by sunlight as we move out of trees and into the open. But that bleaching recedes as our bodies become used to that lighting, producing a level from which the next change will in turn be perceived.

Finally, Wylie also draws out the significance of various technologies and object to such practices of walking. In this case, relatively mundane objects like walking boots mediate encounters with the rough terrain and so expand the range of actions possible for the walking body. However, these boots also act upon that body in ways that may be more negatively affecting (rubbing and blistering skin, for example). Equally, objects/technologies like maps play an important role in (only partially) orientating the walker (see Lorimer and Lund 2008). As Wylie (2005: 241) notes, his:

companions were Ordnance Survey Explorer maps; more than adequate because the Path was usually obvious, and walking it required little skill. In the face of the Path's muscular consciousness, however, the maps' surface tracing of roads, paths and settlements faded from view. This gave way to depth and contour in rust-coloured nests and fans. Studied in the evenings, the maps unfolded fraught topographies...They bulged and rolled with landscape; the coastline in waveform, with whirlpools of gradient, leg-buckling plunges to sea-level and languorous, aching rises.

Together, we gain a sense of the (un)doing of landscape and of self through the practice of walking, and, in this, the focus falls more on landscape in terms of materialities and sensibilities rather than representations, social constructions, or ways of seeing.

Archiving and narrating movements

A prominent feature of work thinking about embodied experiences and the co-emergence of self and landscape through walking has been its auto-ethnographic orientation. Wylie's walks discussed above, and the commentaries on them (see Box 5.3), unfold from the researcher's first-hand experience of walking through these landscapes. This is a prominent feature of a range of work associated with NRTs (see Chapter 7). One point that this raises, though, is the question of writing. One thing we encounter in Wylie's work is a creative attempt to narrate such encounters (also see Rose 2010b). Such a creative-narrated approach is by no means unique to NRTs' engagements with landscape. A concern for narrating relations with landscape can be found in a host of different areas of geographic work since the post-positive turn of the 1970s (Daniels and Lorimer 2012). However, this is something that has commonly been associated with NRTs' approach to landscape.

This concern with the potential and power of narrative accounts extends beyond the realm of auto-ethnographic work, though. This can be seen in at least two senses whereby the narratives provided by others – both deliberately literary but also in the form of diarized accounts – can give an insight into the co-evolution of bodies, selves, and landscapes. For example, Wylie (2012) looks at the work of the author Tim Robinson and his attempts at describing specific landscapes in terms of experience. Walking in the landscape – and specifically, Aran and Connemara's landscapes in the West of Ireland – forms a key part of Robinson's 'material' for his books. For example, his 'Aran' volumes are structured and narrated as extended walks through the landscape during which he encounters locals, visitors, incomers, stone walls, cottages, churches, flora, and fauna. Wylie (2012: 370) identifies a "devotion to the intricacies of local material cultures, and to storehousings of lore, custom and tradition". This walking presents a means of immersion and attention, or in the terms used earlier in this chapter, of dwelling in this landscape. In Robinson's work:

> walking, is seen as proper to dwelling...it prompts reverie, closer attunement between self and landscape, and thus the in situ elaboration of a wider ethos, one in which life is best lived through close synergy with local life forms and rhythms.
>
> (Wylie 2012: 371)

Extending such connections between written accounts, moving through landscapes, and notions of dwelling, Wylie (2002b) looks at the extensively discussed movements of Robert Falcon Scott and Roald Amundsen through the frozen landscapes of Antarctica during their attempts to be the first to reach the South Pole. Here, Wylie (2002b: 250) does not focus on geopolitical matters such as the ideologies motivating the expeditions into such landscapes and rather aims to "discuss these events as events, as an unfolding ensemble of geographically and historically specific practices and subjectivities". Drawing upon a range of written accounts (primarily those found in the diaries of the expeditions), Wylie is interested in the different styles of dwelling in, and moving through, the landscapes present in these two journeys. In this, we find detailed accounts of the agency of the shifting landscapes encountered and the different approaches taken by each party in the face of this; the visual deceptions found in the landscape's featurelessness; the more-than sensory interactions of humans, dogs, excrement, meat, and so on that unfold in navigating this landscape; and, the ultimate fates encountered by each. This all emerged from moving written accounts of moving.

Differently mobile bodies

One important point that has emerged in light of NRTs' engagements with questions of dwelling within and/or moving through landscapes is that of how different bodies might have different capacities to act and, by implication, to be acted upon in such circumstances. Given the phenomenological heritage of much of the work discussed so far in this chapter, and the often personalized, auto-ethnographic narratives that can be found within it, there is a risk that some of the universalizing gestures of past geographic engagements with phenomenology are repeated (or at least perceived) despite the concomitant emphasis on difference, process, and becoming (on humanistic geography's universalizing take of human-landscape relations, see Rose 1993). While NRTs move away from the conception of difference often found in geographic work on identity – thinking more in terms of differences in degree rather than differences in kind – it does seem evident to suggest that a host of broad characteristics might bring different experiences and encounters in and with landscapes. Older bodies might, for example, encounter rural landscapes in different ways than younger bodies (see Maclaren 2018), and bodies of different races might be differently placed within rural landscapes than in urban landscapes (see Tolia-Kelly 2006b). And there is also very likely going to be a great deal of difference within such categories.

The importance of such embodied differences can be seen clearly in Hannah Macpherson's (2008, 2009, 2010) writings on blind and visually impaired people's embodied encounters with landscape. While academic and political debates around such disabilities might push toward the need to 'give voice' to under-represented groups, Macpherson (2010: 3) turned to NRTs in an attempt to gain an "understanding of the ways in which visually impaired visitors both produce and are produced through particular countryside spaces, social scenarios and research settings". In particular, this meant thinking about body–landscape interactions as processual whereby "the body takes shape through its interactions with other objects, bodies and landscapes" (Macpherson 2010: 4).

In exploring these themes, Macpherson acted as a 'sighted guide' and conducted interviews with members of specialist blind and visually impaired walking groups that visit areas of the Lake District and the Peak District of Britain. In particular, Macpherson (2008: 1080)

> aimed to offset ablist understandings of rural landscape based on a normative sense of the individual sighted body, for perhaps nowhere is an 'ocular-centric' approach to knowledge more evident than in the understandings and representations of the British countryside.

From this, Macpherson (2008: 1080) suggests that:

> the presence of people with blindness or visual impairment in areas of scenic landscape disrupts the traditional association of landscape with an individual's visual apprehension and draws attention to the other embodied, tactile, and sonic ways in which we might apprehend the landscape.

Through Macpherson's (2010: 9) attention to the sorts of (often humorous) interactions that took place in these bodies' movements through these landscapes with others (both sighted and unsighted) – conversations along the way, trips and stumbles over rocks and uneven ground, attention to the sounds present or produced in the landscape – what we start to see is the "'inter-corporeal' nature of moving through a landscape as part of a group". Again, there is a co-emerge of bodies, selves, and landscapes in the unfolding of practice rather than any of these being fixed in advance of that unfolding. What these bodies could do – both together and themselves – was not given in advance but rather unfolded in their relations with others and the landscapes they were situated within.

Haunted and spiritual landscapes

The work discussed in this chapter has largely focused on the ways that bodies and landscapes interact in light of the specific capacities of bodies and the material affordances of landscapes. Agency in this has been distributed such that a host of non-human matters – trees, rocks, crops, buildings, paths, and so on – hold capacities to affect (and be affected by) the bodies that move through the landscape. However, there is potentially more going on here than first meets the eye (or foot). One thing that NRTs have drawn attention to are the ways in which "events continue to reverberate in and around places long after they have occurred" (Maddern and Adey 2008: 291). In this, we can experience "the just perceptible, the barely there, the nagging presence of an absence" (Maddern and Adey 2008: 292). This is what Peggy Phelan (1993) has referred to as the 'unmarked', that which is not clearly marked as here but remains palpable in its reverberation in and across space (also see Thrift 2000). In moving through and living in a host of landscapes, we experience traces of "previous [forms of] social life, inhabitants, politics, ways of thinking and being, and modes of experience, all of which may inflect social experiences and translate into new material affordances" (Maddern and Adey 2008: 293). It is by no means easy to represent such potentially unrepresentable hauntings (Holloway and Kneale 2008). This both expands the registers through which bodies and landscapes co-emerge and are co-constituted, and can appear in both banal and quite spectacular settings which variously elude our efforts at capture.

Haunted landscapes

One way that we can see these affective absent-presences which exceed easy representation is in the traces of past lives lived in a variety of landscapes. This can act at and across different scales and registers. These absent-presences can be tied to questions of collective pasts and be widely felt events (tragedy, war, injustice, persecution, and so on) amongst proximate and distant populations (see Till 2005). They can also be quite personal and matter in the context of the ongoing (co-)emergence of individual subjectivities (Wylie 2009). And, they can be both simultaneously, affecting differently and with varying intensities across space and time, individuals and collectives.

While geographers interested in landscape have for some time been concerned with various practices of memorialization, both individual and collective (see Hoelscher and Alderman 2004; Johnson 2004), recent work interested in 'spectral geographies' has shown other senses in which the past can come to matter in and through landscape and specifically how this can lead to "the disruption or dislocation of normalized configurations and affordances of materiality, embodiment and space" (Holloway and Kneale 2008: 303). This has included attention to quite literal senses of haunting and ghosts through attention to their portrayal in media and through practices such as séances (see Holloway 2006; Holloway and Kneale 2008; Kneale 2006). However, this has also meant an attention to how in urban landscapes, for example, we might encounter "apparitions which are the unintended consequences of the complexity of modern cities, cities in which multiple time-spaces are being produced, which overlap, interact, and interfere" (Thrift cited in Holloway and Kneale 2008: 298). We can see this in Tim Edensor's (2008) account of how the urban landscapes of Manchester continue to be haunted by its industrial past. As he explains:

> The speed of social and spatial change throughout the twentieth century means that the contemporary era is the site of numerous hauntings, for the erasure of the past in the quest of the ever-new is usually only partial ... Accordingly the contemporary city is a palimpsest composed of different temporal elements, featuring signs, objects and vaguer traces that rebuke the tendencies to move on and forget.

> (Edensor 2008: 313)

In the case of Manchester, Edensor (2008: 314) shows how a series of attempts to reinvent the city as a cool space for middle-class residents and leisure/consumptions practices means that:

> Swathes of former manufacturing space and lower class housing have been converted into luxury apartments, offices and retail outlets. Scrubbed clean and imbued with design features, the industrial and class histories of such sites are effaced.

However, such reinvention is only ever partial. The landscape retains traces of past lives which haunt and affect those moving through it. The decaying facades of once prominent buildings give clues to good times previously had. As paint peels past functions show through, providing glimpses of, for example, previously covered signs that give indications of changing tastes and dated styles. Spaces once busy and bustling are now subdued. Train lines are left abandoned, past sounds are no longer heard, and energies and atmospheres have long dissipated and leave spaces lifeless. But this can all be sensed, not in the scripting of memory through memorials or monuments but in the uncoordinated circulation of traces of the past that once meant something and still retain the power to evoke something of that past life.

Spiritual landscapes

In addition to these ghosts in and of landscapes' pasts, there are other ways in which absent presences can make themselves felt as bodies live in and move through landscapes. One way that geographers, again working both directly in light of and more broadly in the wake of NRTs, have considered this is in terms of spirituality that might be seen to be bound up with various landscapes (see Dewsbury and Cloke 2009). While there have been a range of quite prominent critiques of the place of religious and spiritual belief in contemporary society, geographers here have recognized that for many people, "spirituality is not incidental to everyday life, but is instead fundamentally constitutive of it" (Yorgason and della Dora 2009: 635). And importantly here, this is not necessarily just about formal or organized religion. Rather, for such geographers, "Spiritual landscapes open out spaces that can be inhabited...in different spiritual registers" (Dewsbury and Cloke 2009: 696). This extends beyond the realms of organized religion into a host of belief systems which matter to peoples' encounters with a range of different landscapes.

A key context in which such spirituality comes to matter, and one which expands on the discussions in this chapter, is pilgrimage. Pilgrimages take many forms and are motivated by a host of spiritual and non-spiritual factors. Commonly, though, pilgrimages involve "a journey to a distant exceptional location, which entails meaningful interactions, authentic experiences and 'extraordinary' encounters" (Scriven 2014: 249). Those encounters might happen at that destination but the journey itself and the landscape moved through as part of this can also matter. This dual sense of pilgrimage – as being about a landscape visited as well as one moved through – is clear in the work of Veronica della Dora and Avril Maddrell on Christian pilgrimage. More specifically, Maddrell and della Dora (2013) compare Ecumenical pilgrimage and Orthodox Christian pilgrimage and draw a distinction between them in terms of the directionality in their engagements with spiritual landscapes. Here, Ecumenical pilgrimage is deemed to be about 'horizontal engagements' with meaningful spiritual landscapes in the form of moving across a landscape which is experienced as a threshold to worship and spiritual encounter (a journey). This means that the destination is less a concern. Rather, the literal (and metaphoric) journey the pilgrim goes on as they move through these landscapes is the central concern. By way of contrast, Orthodox pilgrimage is about 'vertical engagements' with meaningful spiritual landscapes in the form of the veneration of holy icons or relics (an object), often based within a specific self-contained locality (a church). That said, that location is often within a particularly spectacular or inspiring landscape.

Common to both practices here is a specific sort of aesthetic engagement with landscape. As Maddrell and della Dora (2013: 1112) note in relation to these two approaches to pilgrimage, the

> Aesthetic and spiritual experiences are similar in that they entail a sense of wonder caused by a combination of objects, forms, smells, or sounds: the rocky pillars of Meteora abruptly emerging from the flat surface of Thessaly; the tapestry of colours and contrasting textures of woodland, rocks, and sea on the Isle of Man; the smell of incense and chants; the penetrating gaze of an icon, and so on... can elicit a particular sense of wonderment.

The landscape's affective capacities and meaningfulness here extend beyond its specific physicality or morphology. There is something encountered or experienced in these landscapes which matter in their movements through and to them which goes beyond the visible/tangible nature of the landscape encountered.

Looking to certain types of pilgrimage that are premised on a much looser sense of spirituality can draw into ever-sharper relief such embodied, affective dimensions of the pilgrim's co-emergence with such spiritual landscapes. This can be seen in Mitch Rose's (2010b) ethnography of New Age Spiritualism and the subjectifying practices its pilgrims engage in through their engagement with the landscapes of the Giza Plateau. New Age Spiritualism here is a diverse movement with no central belief system, no holy text or dogma, and no formal membership or clergy. One thing that gives it unity, though, is the way in which spirituality is seen to extend throughout human and non-human bodies and matters meaning that landscapes are commonly imbued with significant meaning and energies which might be tapped into and channeled.

Echoing the discussion of ghosts above, there is an evident tension in such New Age Spiritualism's practices of pilgrimage, and the engagement with landscape that unfolds in this, between absence and presence. Rose (2010b: 510) suggests that these tourists "perform earthly rituals – supplications to gods hidden and dispersed through sand and rock, and yet, available in lost recesses and forgotten reserves waiting to be tapped" (Rose 2010b: 510). Here, the pilgrims encounter the pyramids of the Giza Plateau in a very specific way. For them:

> The pyramids were not built as tombs but as spiritual power plants for collecting and dispersing such energies ... spiritual machines: you have to understand both parts, the machine is the nuts and bolts but the spiritual is a part of it, it was a device to produce free energy for all the people.

> (Rose 2010b: 511)

Such energies are taped into through a range of embodied practices and sensing: collective chanting, through the generation of resonances and vibrations, and by touching and feeling features in the landscape. Again, these landscapes are not just inert backdrops to human practices of movement and inhabitation but rather are filled with meanings and energies which shape the identities and sense of self of those who visit and/or move through them, and which are themselves shaped through such practices of movement and inhabitation.

Conclusion

This chapter has introduced NRTs' engagements with landscape and has introduced a range of key themes and ideas that have characterized such engagements. In particular, the chapter has shown the ways these engagements have departed quite significantly from a range of prominent geographic work on landscape. Notable in this is a movement from thinking about landscape in terms of its morphology or as a text or representation to working with landscape as something that is practiced. This has been a contentious move and a host of debate has unfolded, particularly in terms of what aspects of landscape might have fallen out of view. It is evident, though, that NRTs' concerns with the processual nature of practices, embodied experiences, materialities, and affects have led to a particular take on landscape which, in emphasizing the co-constitution of self and landscape, has brought a range of previously under-attended to matters into critical view.

In suggesting this process-based co-constitution of self and landscape marks a key feature of NRTs' engagement with landscape, this chapter has introduced a host of ideas through which this has been understood. Prominent here are ideas around dwelling and how we might understand landscapes less as something upon which human activities become marked and rather something with which people live on varied timescales. Ideas around dwelling have been worked with and developed in an effort to rethink the sorts of binary logics that might be present in how we think about a body's relations with a landscape. Distinctions between self and world become blurred as each side of such binaries becomes inherently entwined and co-constitutive. Critically, though, geographers working in light of NRTs have sought to move away from the potential risk of romanticism present in ideas such as dwelling where we end up with 'proper' or 'authentic' ways of living with landscape and, by implication, 'improper' or 'inauthentic' ways of living. That might be related to the specific means through which the relation unfolds (or does not unfold) – with attention to practices like walking or looking – or in terms of assumptions made about the specific bodies that engage (or do not engage) in such relations. As part of this, we saw a movement toward concerns with openness, unfolding, dynamism, differentiation, and engagements which unfold on more immediate as well as longer-term durations between self and landscape. Dwelling not only has come to be seen here as unfolding through movement but as a movement itself.

Finally, this chapter has shown how such an approach to landscape extends the senses in which landscapes become constituted and meaningful. These ideas present a host of opportunities to rethink the relationship between memory and landscape but also to extend the sorts of meaningfulness and sensing that might be recognized as unfolding in people's relations with landscapes. Notions of haunting have shown clearly how the past matters to encounters with landscape outside of the realm of formal memorialization or things that can be obviously read from the landscape. Absent-presences haunt landscapes and affect those who inhabit and move through them. Equally, broader notions of spirituality have been shown to matter to how people move through and engage with specific meaningful landscapes. Affecting meanings and energies circulate through and are potentially tapped into by those engaged in practices like pilgrimage.

It is important to note that these developments do not necessarily mean that a new norm has been established in terms of geography's broad and ongoing engagement with landscape. It is not necessarily the case that matters relating to landscape's interpretation no longer matter in terms of the position of a host of people in relation to them. Rather, the work discussed in this chapter variously poses challenges in terms of thinking about the registers upon which such meanings might be made, the ongoing nature of such making, and the range of affecting materialities (and absences) which contribute to this ongoing landscaping and selving.

Further reading

Wylie (2007a) provides an introduction to geography's engagement with landscape and how this has been approached in a range of ways over the course of much of the 20th and early 21st centuries. This includes discussions of work influenced by NRTs, showing their concern with embodied experience and practice as well as a range of further trajectories emerging from such work around affect and vitalism. **Howard et al. (2018)** provide an extensive introduction to the way landscape has been studied across the humanities and social sciences. This extends well beyond the realms of the impact NRTs have had on the study of landscape. Of particular relevance here though are chapters that provide an introduction to: more-than representational landscapes (chapter 7), landscape and phenomenology (chapter 10), and landscape and sound/music (chapter 21). **Macpherson (2010)** provides an accessible introduction to some of NRTs' general ideas and how these have translated into work on landscape. Macpherson reflects on the methodological challenges of researching landscape in terms of embodiment and what we really mean by bodies here. On the former, this is not so much an argument for or against specific methods but more a recognition of the challenges present and how, at a general level, we might respond to them. On the latter, Macpherson raises questions around bodily differences, particularly in terms of disability, and how they might matter when it comes to thinking about body–landscape relations.

6

NON-REPRESENTATIONAL
THEORIES AND PERFORMANCE

Introduction

Returning to the scene this book opened with, we encountered a number of bodies moving through a city on the morning after the night before. Some of these bodies, given the timing within the UK academic year, will likely be feeling the effects and affects of a night out. Some of these bodies will have been in the bars that surround the area they are walking through. Others might have ventured further to some of the city's nightclubs. Others still might have been in the Student Union. Whatever the case, it's likely that these bodies encountered a combination of sound and movement which inflected the unfolding of that night out.

When I was an undergraduate student at the University of Glasgow, I worked in one of the Student Unions as a trainee sound and lighting engineer. Once a week, the Union would host local unsigned bands and I would help the Union's sound engineer set up the stage, test the equipment, program the lighting, and sound check the bands. Setting up unfolded during the late afternoon when this part of the venue was partly open to students to eat, drink, and socialize in. The room would be fully lit, and so the stark, sticky, and stained black coverings of the stage and its surrounds were visible. A not so faint smell of bleach often filled the air. Each night, a different themed night took place here: 'Revolution' (a rock night) on Tuesday and 'Cheesy Pop' on a Friday were commonly the busiest. Anyone who has ever seen a night club with the lights up will have a sense of the uninspiring nature of this space.

For this night, each band was tasked with selling tickets, and this determined the order in which they performed. The venue was quite large so this rarely translated into anything approaching a full-house – including a seated balcony, the capacity was something like 1,000. The band that sold the most would 'headline' (go last), the band who sold the least tickets would go first, and others would be positioned accordingly in between. Normally,

I'd mix sound for the first band or two and run a simple, largely automated stage lighting program before the sound engineer took over sound for the final band where I would do a more elaborate lighting set, trying to sync the movement of the lights with the tempo and dynamics of the music I was hearing for the first time.

As would be expected, there was variability here in terms of the quality of the bands playing and the size of the audience they drew. At times, if the bar staff and I were excluded, the band members on stage would approach the numbers watching/listening. Normally huddled somewhere toward the back of the venue, a small number of friends would stand awkwardly, their clapping at the end of songs competing with the hum of the PA system. Other times, though, something quite different happened. Getting on for 20 years later, one band still stands out in my memory. Somewhat unusual in style for the time – most of the bands hosted were a close variation of early 21st century indie-rock – this band played (without irony) a mix of 70s hard rock and 80s 'hair metal'. They had brought with them a sizable and enthusiastic following who stood prominently at the front of the stage. Screamed (but still melodic) vocals dovetailed with technically impressive guitar solos, delivered with an appropriate dose of posturing – one foot on the monitor speaker and well-practiced 'guitar-face'.

Something was happening in this performance. As they had sold a good amount of tickets, they featured later in the billing. This meant that the sound engineer was mixing their sound and I was doing lighting. As a guitarist myself, I found myself following closely the guitarist and both the riffs the songs were base around and the solos he took, getting a sense for how they built and some trends that seemed to be occurring in terms of dynamics and length. I responded by trying to choreograph the lighting to complement this. For the riffs, I tried to sync the movements of the lights to their rhythms and accents. For the solos, I experimented with training a spotlight on the guitarist at the start of the solo, adding moving spots as the solo built, then moving to full strobe-flashing as the solo reached its crescendo before killing the lights completely as the solo and song ended. After a song or two we seemed to be in sync and I noticed that the guitarist had spotted this too – at the end of the second song where I had tried this, I caught him looking over to our booth, smiling. In these moments, with metal-heads moshing and charging around the dance floor, lights strobing, legato runs speeding up and down the guitar's neck, this space became something quite different – the bleachy smells, awkward claps, and sticky floors of earlier faded into the background. Instead, some kind of collective affective entrainment took place with sound, light, and bodies all caught up in the movement.

This again might seem like an odd anecdote for a geographer to be writing about. However, a key focus for geographers engaging with NRTs since their inception has been a concern for creative and artistic practices and the spaces-times of experience these both happen within and act to create (McCormack 2008b; Thrift 1997, 2000, 2003). The metaphor of performance played a prominent part in the initial articulation of NRTs, it being used in the context of actual artistic performance as well as more broadly in reference to how life can be seen 'as' performance (see Chapter 1). In this chapter, the focus will largely stick to the former, to artistic forms of performance. Artistic performances, "as a mode of embodied activity that transgresses, resists, or challenges social structures" (McKenzie 1998: 218), maintain within them the possibility of liminality. It has become evident from a host of recent scholarship that various types of creative practice and artistic performance hold the potential to shape how people relate to each other and how they experience specific places, changing how they are experienced and their meaning (even if temporarily).

In setting the scene for a discussion of such an interest in performance, the next section will give a brief introduction to geography's recent interest in the performing arts both in terms of their theories and their practice. From there, the chapter will look further at what Thrift (2008: 135) has called "the art of (and the art of valuing) the now". In doing so, this chapter will both introduce this turn toward performance in geography and the various sources of inspiration for this, before specifically focusing on NRTs' interests in dance and the generative relations of body and space that this has been seen to entail. Further, and looking to the performance of music as an 'art of the now', this chapter will discuss what Nichola Wood and others have called 'musicking' (Wood et al. 2007). Next, the chapter will turn to questions of power and performances. In particular, this will look to NRTs' concerns with issues of power in performance, specifically through a discussion of Derek McCormack's writing on power and rhythm in the context of eurhythmics (McCormack 2005). From this, the chapter discusses the matter of situating performances. This has been a recurrent issue throughout this book's discussion of NRTs and has arisen in a range of contexts (practice, landscape, affect, etc.). In reflecting on this, the chapter concludes by comparing and contrasting the approaches of a range of geographers interested in dance and music who variously juxtapose or hold in tension questions around 'the representational' and 'the non-representational'.

Performance geographies

As Amanda Rogers (2012a: 60) suggests, recently, "cultural geography has witnessed a burgeoning engagement with the theories and practices of the performing arts" (also see Abrahamsson and Abrahamsson 2007). NRTs have played a significant part in the development of that interest. This can be seen, for example, in the focus of early themed issues of journals on topics such as 'The possibilities of performance' which talk broadly in terms of practice and performance (see Latham 2003b) as well as Thrift and others' attention to specific performance genres (Dewsbury 2000; McCormack 2005; Thrift 1997). However, geographic interest in the performing arts has by no means been limited to work associated with NRTs, both given past engagements in geography and in recent developments which extend across a host of disciplines. In the past, geographers' interest in performance might have involved more metaphorical discussions of everyday life being a performance or the discussion of a host of phenomena through dramaturgical metaphors. For example, peoples' routine movements through space have been described as 'place-ballets' (see Seamon 1980), and the language of 'the performance of the self' has been used in reflecting on individuals undertaking specific roles in certain situations (Crang 1994). More recently, there have been a number of examinations and calls for an attention to the "deeply social character of coded performances of identity", for example, on the basis of gender, race, class, nationality and/or ethnicity, when it comes to the performing arts like dance (Nash 2000: 657; also see Leonard 2005; Noxolo 2018). This becomes very clear when it comes to thinking about specific dance traditions (for example, on rumba see Hensley 2010, and on tango see Nash 2000) or in the efforts of geographers to create performance-based outputs from their research with specific social groups (see Johnston and Pratt 2010).

Rogers (2012a: 60) suggests that such recent work engaged with performance (again, largely in terms of artistic practices) has allowed for the recognition that:

> whilst the performing arts reveal the experiential qualities of space and place [as suggested by NRTs], they also provide a way to think about their power-laden politics. Performances can reflect contestations around place and identity, they highlight how people and places are embroiled in flows of capital, and they can even intervene in the construction of built, material environments.

While such work is evidently different in orientation to the work that this chapter will focus on, this should not be read as suggesting that NRTs' interests in performance are apolitical. The attention here to the experiential qualities of space and place is not something that is necessarily separate from power-laden politics. To return to a theme raised more than once so far in this book, it is rather the case that NRTs' here think politics differently. There are many ways that we might think about and foreground the relations of the corporeal and the cultural (McCormack 2008b). The effort here is actually to make things 'more political' by "expanding the existing pool of alternatives and corresponding forms of dissent" (Thrift 2003: 2021). Part of the point is to recognize the ways in which politics might be expanded through an engagement with debates, ideas, and practices that operate outside of the norms of social science subjects like geography.

Such points aside, it's important to recognize here that this is not necessarily a one-way conversation from the performing arts (as a collection of practices and an academic arena) to geography. There has also been conversation amongst scholars from these various academic contexts when it comes to broadly geographic themes such as landscape, ecology, place, and urban environments. As Rogers (2012a: 64) surmises:

> Academics in geography and performance studies ... share similar concerns around how landscape can be seen as performance, the relationship between performance and environmental change, what performance can achieve ecologically, if being overtly activist can alienate audiences, and whether performance can raise consciousness or change behavior.

Rogers (2012a) shows in more detail than is possible here that there is a growing body of scholarship whereby theatre scholars and practitioners, performance scholars and practitioners, geographers and others are engaging in dialog around such themes/settings in ways that are geographic in nature (thinking about spatiality, site-specific performance, and so on) but which are potentially novel in their format, presentation, research practice, and so the ultimate direction that this work takes. Mixing insights from performance and theatre studies (Pearson 2006; Schechner 2002, 2003), anthropology and phenomenology (Ingold 2000; see Chapter 5), as well as critical dispositions from a host of environmental and political projects (Kershaw 2007), such work leads to a position whereby discussions circulate around the inherent interconnections of embodied experience, cultural traditions, certain socio-political contexts, and, importantly, the interlacing of these with the physical environment in which those performances unfold. This performance work is very much concerned, in quite broad terms, with "the affective ties between people and place" (Pearson 2006: 4).

Arts of the now

Moving more specifically to NRTs' engagements with performance, in referring to performances as 'arts of the now', Thrift (2008) draws on a host of performance scholars who have sought to emphasize the immediacy and presence in/of performances and the improvised nature of their unfolding. During the 1990s, a significant debate unfolded in performance studies around issues of mediation, repetition, and 'liveness' in performance, much of which questioned the impact that various forms of mediated reproduction had on performance. There was concern here that media-based re-presentations of performances (via television, film, or now the internet) acted to dull the liveliness of performance in removing the audience from the actual happening of the performance event (see Auslander 1999). The argument here was that:

> Performance's only life is the present. Performance cannot be used, recorded, documented or otherwise participate in the circulation of representations of representations; once it does so it becomes something other than performance. To the degree that performance attempts to enter the economy of reproduction it betrays and lessens the promise of its own ontology. Performance's being becomes itself through disappearance.
>
> (Phelan cited in Thrift 2008: 135).

There is a clear affinity here in the reference to representation's limits with NRTs' suspicions around viewing the world 'as text' (Dewsbury *et al.* 2002; Thrift 1996). While a performance might emerge from or include a whole host of texts (scores, charts, scripts, stage directions, and so on), it is not reducible to such texts. An important reference point here for Thrift was Isabela Dora's suggestion that: "If I could tell you what it meant, there would be no point in dancing it" (cited in Thrift 1997: 139). There is something ephemeral and performative in the happening of performances here that goes beyond texts and which contributes to a particular sense of space and time (see Box 6.1). These practices are expressive rather than simply reflective of some prior meaning that might be found in one of the texts which accompany, direct, or document the performance.

In articulating how performances can be seen to be 'arts of the now', Thrift (2008: 135–137) specifies six defining characteristics. Here, performance is:

- something that heightens everyday behaviors rather than being separate from them;
- liminal in the way that it challenges, plays with, suspends, and potentially transforms social structures and norms;
- concerned with constructing 'unstable times' in its status as an event with beginnings and ends that reach beyond its specific unfolding;
- concerned with constructing 'unstable spaces' of possibility that are fleeting, risky, interactive, and that might go wrong;
- potentially normative in its reiteration of certain ideas, norms, and behaviors;
- hard to capture, 'mark', reproduce; is unsayable and unstable.

BOX 6.1 THE PRODUCTION OF SPACE-TIME?

The idea that space is not an inert or neutral container for human action and instead is something that is produced through actions such as those of a performance is by no means unique to NRTs. For example, geographers have for some time engaged with the writings of Henri Lefebvre on the production of space (and more recently rhythm) in thinking about how spaces are made in and by specific societies/economies (see Elden 2004b; Mitchell 2003). Lefebvre's (1991) 'trilectic' of social space has been prominent here. This entails a concern for the interrelations between representations of space (plans, policies, regulations, etc.), spatial practices (the routine things that people do day in, day out which largely conform to such representations), and spaces of representation (where moments emerge where change might take place), which all coexist with varying prominence and import at any given time.

However, while this might sound similar to NRTs' discussion of the generative nature of performance, NRTs tend to look at such production in less structured, more messy ways where it isn't always clear where the boundaries between spatial practice, representations of space, and spaces of representation might be drawn. The influence of post-structuralist and vitalist ideas here is evident in the emphasis on emergence, becoming, multiple sources of agency and activity, and so on. In this, there is not the same sort of Marxist, normative line present as in Lefebvre's work whereby representations of space are assumed to alienate, for example, as an exercise of power over a body. In work such as Lefebvre's, even if by implication, there is a prior qualification given to various aspects of the trilectic (positive or negative). Each component is benchmarked according to a preexisting set of criteria aligned with his broadly Marxist agendas in producing a 'critique of everyday life' (see Simpson 2008).

Rather than Marxist social theories/philosophies, then, much of the influences on how NRTs have approached the production of space-times through creative acts has come from performance studies and the creative arts (as well as broader philosophical discussions around process, becoming, affects, and the like). These do not necessarily pre-judge how a performance will function (or malfunction). Instead, attention is given to the unfolding of the performance event itself. While representations might exist around a performance, these are by no means necessarily going to affect things negatively. A score presents a starting point from which expressive potential might (or might not) emerge in the course of a performance, for example. A host of relations can unfold here between bodies whose capacities to affect and be affected are not yet known to us (also see McCormack 2005). Constraints can potentially enhance a body's capacities to affect and be affected as well as diminish them. The interesting thing here is how various constituents involved in a performance work across such boundaries, boxes, and qualifications; or how something might come from outside of them in surprising and unsettling ways.

There is a danger that much of this characterization misses the great diversity in what form performances can take and over-emphasizes certain aspects of illusiveness, liveliness, criticality, and so on compared with the single reference to their potential normative potential. There is a danger also that this moves in a direction which doesn't quite realize the breadth of performance styles, extending across "the accessible to the avant-garde" (McCormack 2008b: 1823). There's a lot going on here beside performances' potential to create in liminal ways. This is not necessarily the only aspect of performances that might matter or be important in a given situation. However, such themes clearly raise potentials for consideration and pose challenges to many geographic engagements with performance as a geographic concern (see Brigstocke 2014).

Experimenting with dance

One 'art of the now' that Thrift initially focused on in his discussion of NRTs was dance. Dewsbury (2011a: 64) suggests that dance is an apt starting point as "dance is the art form least reliant on representation as it primarily works with and demands space or spacing alone". However, when it comes to NRTs' initial experiments with dance, these operated less at the level of actually researching specific genres of dance or spaces where dancing happens and more at the level of what inspiration might be taken from dance as a practice, what it might tell geographers about how the world works, or proposals for themes that might be researched in subsequent work. For example, Thrift (2008) identifies three key 'expressive potentials' in dance as follows:

- *Dance and the body*: For Thrift (2008: 140)

> dance can sensitize us to the bodily sensorium of a culture, to touch, force, tension, weight, shape, tempo, phrasing, intervalation, even

coalescence, to the serial mimesis of not quite a copy through which we are reconstituted moment by moment.

Dance is suggested to provide a route into how cultures are (re)produced through embodied practices. However, Thrift (2008: 141) also draws attention to how: "dance can produce new bodily expressions which turn on the body's power to purposely transgress, play, or dissimulate. The body is not just written upon. It writes as well". Thrift (1997) draws a distinction here between dance being less self-evidently about discourses of power and control and more about play or experiment; that it is about "off-balancing, loosening, bending, twisting, reconfiguring, and transforming the permeating, eruptive/disruptive energy and mood below, behind and to the side of focused attention" (Schechner cited in Thrift 1997: 145). Thrift (2008) connects these points to specific forms of improvised dance. However, more generally, what we have then is another route for NRTs into the ways in which norms are both (re)produced but also played with through the unfolding of specific (creative) practices whether deliberately based on improvisation or in light of the differences that appear within repetitions of something that might be heavily choreographed.

- *Forging identity*: Developing these ideas about the embodied nature of dance and the performative nature of this, Thrift also discusses dance as being explicitly about the forging of identities. This is pitched more at the level of momentary forms of subjectification than a more formalized sense of identity that might be associated with broader social or geographical contexts. The focus is on "configuring alternative ways of being" (Thrift 1997: 147). Here, Thrift suggests that such identities (or subjectivities) formed through dance can be based on something as simple as dance 'evoking a mood'. This means that "Identity can be constructed by dance at the level of individual experience, or at the level of social assemblages" (Thrift 2008: 142). What we have here is a concern for how new forms of (self-)awareness and persistence might emerge through the practice of dance, how desires and energies might be channeled, and, with this, how dancing might lead to individuals making sense of the situations in which they find themselves.

- *Dance and city*: Thrift suggests that studying dance can help geographers understand urban life. Dance here becomes something more metaphorical given the way that connections are made, for example, to ideas around 'rhythmanalysis' (see Chapter 2), and the sorts of dispositions that would be necessary for understanding the complex choreography that patterns urban living; the movements of bodies, vehicles, goods, information, and so on across a myriad of infrastructures (roads, paths, pipes, cables) and means (in cars, vans, buses, lorries, or self-propelled), in ways that are variously synchronized or syncopated. Or in other words, dance

> enables us to rediscover and rework the plural, performative skills of the city, stimulating both a greater sense of extant situations, and a glimpse of new styles of urban living which might simultaneously produce new senses of how the world is.
>
> (Thrift 2008: 145).

Moving geographies

Moving to work exploring the generative potential of dance more clearly in relation to actual dance practices, Derek McCormack's work has shown how the dual sense in which bodies move – both in terms of the actual movements of bodies in space but also in affective and kinaesthetic senses – can produce or generate spaces (see McCormack 2003, 2008b). This can be seen, for example, in his discussion of 'The 5 Rhythms', a somatic movement practice that attempts to facilitate expressive forms of movement (McCormack 2002). 'The 5 Rhythms' is described as:

> a dance form created to give the space and permission for people to move freely. This easy and effective practice is a moving meditation that brings us home to our bodies, gives expression to our emotions and catalyses our innate capacity to dance.
>
> (Advert for 'The 5 Rhythms' quoted in McCormack 2002: 469).

'The 5 Rhythms' was developed by the American dancer Gabrielle Roth who suggests that these rhythms offer maps "that can catalyse the expressive potential of the moving body" (McCormack 2002: 471). The five rhythms here are 'flowing' (being fluid, flexible, and loose), 'staccato' (being angular, linear, and edgy), 'Chaos' (shifting weight as a result of the collision of the flowing and the staccato), 'lyrical' (moving more lightly), and 'stillness' (the space between beats) (McCormack 2002). In their combination, these rhythmic movements are intended to be transformative and are connected by Roth to a host of spiritual concerns around self-healing. Such concerns could be critically deconstructed in a host of ways, for example, through connecting them to the proliferation of a range of commercialized 'techniques of the self' which respond to a host of contemporary social, economic, and existential crises of identity and purpose. However, McCormack (2002: 473) suggests that such a deconstruction "short circuits the creative potential of the practice by reducing it to a symptom of wider shifts in cultural and political economies". There is also, then, the potential that such practices might "provide new modes of ethical and aesthetic inhabitation" meaning we might want to approach them with a more generous disposition (McCormack 2002: 473).

McCormack's discussion of 'The 5 Rhythms' is deliberately playful in the performative mode of writing it adopts. This seeks to respond to questions such as:

> How does one give a word to a wordless movement without stifling the life of that movement? How, when such movement is often below the cognitive threshold of representational awareness that defines what is admitted into serious research, does one give a word to a movement without seeking to represent it?
>
> (McCormack 2002: 470)

In doing so, McCormack moves off from traditional forms of academic writing in trying to give a sense of, or some consistency to, the expressive movement and fluidity involved in the unfolding of the practice itself. As such, we encounter the practice of 'The 5 Rhythms' through accounts like:

> One ventures from home on the thread of a tune, along sonorous, gestural, motor lines, lines now warmed up, becoming familiar with movements that seem to come from nowhere, yet without thinking, taking hints from others, as a voice invites a **moving into the first rhythm, flowing, the rhythm of curves, circles**, and one begins looking for the security of a phrase, a gesture, a line, something within and along which to move, before becoming more adventurous, as that phrase, that gesture, that line becomes something other by moving through me, moving beyond what it was, through moving something of mine beyond me, and now a track change, and music comes with curves that circle around, picking up and folding into the speeds and directions of another phrase, another gesture, another line, here and there, until before long a roomful of bodies becomes a multitude of curves, each of which is enhancing, inhaling, rising, expanding, and opening, into the curves of another...
>
> (McCormack 2002: 475)

and diagrams that express something of these movements (see Figure 6.1). In an attempt to give consistency to these creative potentials, McCormack draws on Deleuze and Guattari's (2004) account of 'the refrain' (see Box 6.2). The central idea here is that the rhythmic repetition of some type of motif

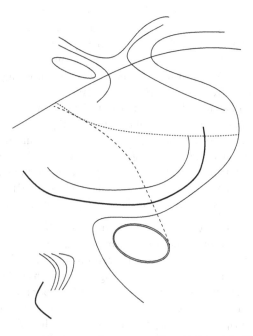

Figure 6.1 Diagramming rhythms and movement (*From McCormack 2002*).

BOX 6.2 OF THE REFRAIN

The refrain is a concept most commonly associated with the writings of Deleuze and Guattari (2004). In music, a refrain is a recurring theme that appears at various points throughout a song. Deleuze and Guattari provide a range of examples of what such a refrain might be outside of the context of a musical performance. It might be a tune tweeted by a bird, a melody sung quietly by a child, and so on. What Deleuze and Guattari see as common in such repeating melodies is that they allow for the unfolding of activity in a given space at that time. They potentially provide something familiar that comforts and so settles that body in that context and so allow it to carry on. Such refrains might act "like a rough sketch of a calming and stabilizing, calm and stable, center in the heart of chaos" (Deleuze and Guattari 2004: 343). They can make us feel at home. There are often a range of familiar sounds that we associate with being at home (the background noise of a radio, for example) which are as much taken note of when

(Continued)

they are absent as when they are present given the loss of comfort that can bring. But that's not just about being settled. Such familiarity provides us with a basis to move off, to do something different, to approach something that might otherwise be disconcerting. Think of a child who is afraid of the dark singing a familiar song, perhaps sung to them by a parent or heard in a favorite TV program, so as to settle themselves into sleep (see Ash and Simpson 2019).

Refrains can also take forms other than those of sound and melody. In this sense, the refrain is "any aggregate of matters of expression that draws a territory and develops into territorial motifs and landscapes" (Deleuze and Guattari 2004: 356 [emphasis in original]). Expression here means something quite specific. As Lingis (2004: 274) notes:

> Expressions are gestures, fancies, rituals, vocalizations, and speech. Expressions never simply represent, depict, describe, or report on the things we couple up with. Expressions anticipate couplings our bodies will make, go back to couplings our bodies have already made and no longer make, slow down the couplings our bodies are making with things and events or accelerate them, detach them or unite them, map them or segment them. Our bodies, for their part, with the couplings they make with things, expose themselves, extend themselves in expression.

Refrains can be things that are seen; they can be found in a gesture or a movement, or anything really that, through its repetition, expresses something that orientates and situates us in these ways, that "are used for organizing a space" (Deleuze and Guattari 2004: 343).

can give consistency or familiarity to the relations that unfold in the course of movement/dance. It is not the case that continuity is produced here by the refrain. Rather, it is about consistency; the holding together of things (even if temporarily) without necessarily resolving issues of difference or dynamism in what is held together. In McCormack's (2002: 476, and again on p 479, 483, 484) account, we regularly encounter the refrain that we might "*Venture from home on the thread of a tune, along sonorous gestural, motor lines*", each time varying in where it leads as he moves through the five rhythms of this practice. In terms of such refrains that unfold in the act of moving itself, McCormack (2002: 478) suggests that:

> one finds oneself ... experimenting with a particular gestural curve, exploring its pathways, its limits, its speeds, until the curve

becomes the movement itself, becomes a refrain, a style that has durational intensity for who knows how long, until from somewhere else another possibility arises, and the affective pathway of another curve takes over. Each rhythm has its own such possibilities, and at some point, these begin moving between rhythms, as one begins experimenting with staccato-chaos, or flowing-chaos. Hybrid, bridging rhythms emerge, passing from one territory to another through the refraining movement of a hand, an arm, a leg, a hip. Returning time and time again, session after session, night after night, the gestural refrains of each rhythmic territory emerge without effort, without the act of thinking what should one do. But this takes time. And sometimes it does not work, and then everything falls flat on its face in a bundle of self-consciousness tied loosely together with the question – what am I doing here?

It's clear that these various refrains – those written and those of movements – produce something like an orientation in these spaces at these times. They allow the moving body to orient itself and provide a basis for something to happen, for variations to emerge in repetition, and so for different sorts of (positive or negative) encounters to unfold. This is not something fixed or necessarily something shared – it's not as grand as something like a 'sense of place' – but rather a sense of this space that holds for a given time before resolving into something else.

Mediated moving bodies

One thing that becomes clear here in McCormack's (2008b: 1827–1828) discussion of dance and movement practices is that:

> Regardless of what kind of dancing might have been taking place, the affective quality of the space in which bodies move is never only something personal – it is a product of a complex mix between music (although music is not necessary for dance), light, sound, bodies, gesture, and, in some cases, psychoactive substances of various kinds.

Thinking about dance and movement with NRTs draws attention to the more-than human set of relations and materialities that play a part in the unfolding of any given circumstance. Agency is distributed across both human and more-than human bodies, amongst vague feelings (both individual and shared) as well as more easily demarcated and delimited objects, amongst that which is present in the moment and that which has happened in the past that might have disposed various bodies here.

One thing worth noting, though, is that McCormack's work isn't just based on auto-ethnographic engagements with dance practice. There are other ways that NRTs have experimented with dance as an 'art of the now'. Again, showing that NRTs are interested in representations in terms of what they do, we can think about "how moving images of moving bodies allow rhythms, movements and affects to emerge, circulate and travel across time and space" (McCormack 2008: 1832; also see Latham and McCormack 2009). Moving images have a non-representational dimension, an affective force, in that they can play a part in the "generation of experience" in being "more than merely symbols" to be reflected upon (McCormack 2013: 143). Such images have "the capacity to produce or catalyze affective transformations in the bodies of those who inhabit its scene" (McCormack 2013: 147). To return to the debates from performance studies about 'liveness', it is clear that while such moving images of moving might be different to a live performance by moving bodies, that is not to say that each showing of such movement doesn't do something for those who witness it.

In showing the affective potential of moving images of moving bodies, McCormack discussed the example of the music video 'Here It Goes Again' by the band OK GO. This video features the members of the band performing a heavily choreographed dance routine on exercise treadmills (see https://youtu.be/dTAAsCNK7RA). For McCormack (2013: 154), this video "has the capacity to move, to transport, by modifying the affective spacetimes of bodies". Part of this comes from the juxtaposition of dancing bodies and treadmills in this video. Treadmills are more commonly associated with the control and regulation of bodies – they regulate the speed and style with which bodies move (without actually moving) for very specific ends. Bodies moving on treadmills are most commonly bodies pursuing a particular project of improved health and fitness. As McCormack (2013: 154) suggests, treadmills "foreground the embodied self as a project motivated by the dream of a healthier, happier, more well-toned you". These are not normally understood to be technologies through which bodies might be expressive in a way that exceeds a very clear vision of what they are to be used for and how they are to be used. As such, for McCormack (2013: 155), the use of these treadmills in this music video:

> seems a remarkably inventive way of choreographing bodies using a technology often associated with repetitive, disciplined movement: an exuberant, if also gently subversive, take on the imperative to become fit through the development of intimate relations with technology.

Importantly here, McCormack recognizes that this video has this affective capacity both during its three and a half minutes of play but also in ways that resound beyond that in a variety of (re)playings. That in part comes with the way it might hold attention, build interest, lead to re-viewing it. Its exuberance is affective and engaging again and again. Equally, though, this also becomes apparent in the way that such moving images of moving bodies circulate and come to be reproduced or appropriated in a host of settings from exercise videos to advertising campaigns. In this way, such moving images of moving bodies "circulate in a wider assemblage of techniques and technologies designed to produce distinctive spacetimes of experience in contemporary life" (McCormack 2013: 156).

'Musicking' and listening

Music comes a close second when it comes to the sorts of artistic practices which geographers have approached as 'arts of the now' in light of NRTs. There has been a fairly consistent, if relatively marginal, interest in music (and sound) amongst geographers for some time, emerging with work grounded in Berkeley School Cultural Geography and carrying on through work informed by New Cultural Geography (and others) (see Carney 1998, 2003; Connell and Gibson 2003; Ford 1971; Leyshon et al. 1998; Matless 2005; Revill 2000a). This has remained relatively marginal given the prominence of 'the visual' to much geographic work. As Smith (1994: 232) noted some time ago now:

> in social science generally, the 'ideology of the visual' has afforded an epistemological privilege to sight over hearing, even though sound (precisely because it lacks the concreteness that so appeals to empiricism) is more allied than vision to those emotional or intuitive qualities on which the interpretive project lies.

The 'emotional' and 'intuitive' qualities that Smith highlights here are important to recent work on music in geography taking place alongside the development of NRTs. To put it crudely, there has been a shift from looking at music as a cultural object or text that tells us something about people and their relations to the places they live, toward an increasing recognition that music is something that is practiced in an experiential sense. There is a live(li)ness to music that goes beyond the dots that might appear on a score, the written words that record the lyrics, or the ways that music might circulate as physical or digital media as part of various national and local networks of economic and cultural exchange (see Finn 2011). The sound of music itself has its own sort of materiality: "the movement of sound waves through the air, completely enveloping us, penetrate our eardrums; the movement of bodies through space, striking objects, setting sound waves in motion" (Finn 2011: 2). It is this live(li)ness that has been brought to the fore by recent work engaged with NRTs.

Spaces of musical performance

A key concern that emerges here has been to focus on the sorts of "environments in which music and sound take place" (Anderson et al. 2005: 640). One antecedent to such a concern can be seen in early discussions of 'the soundscape' (Smith 1994). As Saldanha (2009: 236) notes:

> Soundscape is a crucial term to conceive how sound gives meaning to spaces and places. By filling not only time but space, sounds such as music help orientate the lives of all humans, as much as sight does.

A key distinction emerges here in the emphasis on sound 'giving meaning' to spaces (see Anderson et al. 2005). There are other ways in which music and sound can play a part in the production of space and the sorts of embodied experiences that unfold in and from that. As Feigenbaum and Kanngieser (2015: 82) note, "Sound creates atmospheres through its pitches, tones, volumes, frequencies and rhythms, which penetrate and travel through material and immaterial matter across distances, filling spaces within and between bodies". What we have then is a focusing of concern on "the embodied and lived registers of experience which unfold in the process of making or listening to music" (Anderson et al. 2005: 641).

Such embodied experiences of performances are captured by Wood et al.'s (2007) deployment of the neologism 'musicking'. This term is intended to draw attention to how it is "the materials, meanings, production, experience, and doing of music that matters" (Wood et al. 2007: 869). Again, music here is evidently more than a series of dots on a score and is rather something that takes place in place amongst a range of human and non-human bodies/objects. In this sense, musicking is about any sort of participation in the event of music's taking place, whether that be performing, hearing, practicing, composing, dancing, and so on. In thinking about such musicking, Wood et al. (2007: 869) draw attention to the music's 'materialization' in place, to how "The physical space for music making ...[is]... a thing that is performed and so always in the making". This is not the simple transformation of a blank space into a performance space. Rather, it is recognized here that there are a multitude of factors (always already) at play, be that related to the social circumstances that surround that performance location and those who might (not) attend, the practices of production and promotion that might lie behind the performance-event, or the acoustic properties and potentials of the space where the performance will take place. Further, the degree to which transformations succeed in their aim – in producing a particularly affective, resonant, or meaningful space-time of encounter for those present (both performer and audience) – is by no means certain or felt the same for all involved (see Wood and Smith 2004). Attending to this musicking poses a range of challenges. Its complexity, its liveness, the ephemerality of what is produced, and its largely non-verbal and not easily articulated nature produce a whole host of affects and feelings for those involved (also see Wood 2012).

Such musicking is something that I have approached in my own work in the context of the practice of busking (as well as other types of non-musical street-based performance (see Simpson 2011a)). Busking is usually defined as the performance of music in public spaces such as streets or squares with the aim of soliciting voluntary donations from passersby. Various forms of busking/street performance unfold as part of the everyday life of many cities. I was interested in such types of performance given the way that they specifically draw attention to the relations between the performer, audience, and the environment in which the performance takes place. The street here is significant as

> It is nearly impossible... to separate street performance from the urban environment, for the city exerts a primary influence on both its perception and reception. The shape, texture, and uses of urban space determine behaviours and expectations, performance structures, and the theatrical frame. The width of the sidewalk or shade from a tree, the noise surrounding the performance space, the proximity of other performers, the social as well as the atmospheric climates, the civic regulations concerning performance activities – all are part of the performer's daily, even minute-to-minute negotiations with a fluid and vital urban environment.
>
> (Harrison-Pepper 1990: xv)

Rather than these performances taking place in a venue where a performance might be expected and planning for – where, for example, a stage might be present that differentiates the spaces of performer and audience – the street here must be temporarily made into a performance space. And part of the challenge in that is that the street is not a blank slate to be transformed; it is an existing and unfolding circumstance in which performances must both situate themselves and try to intervene (see Simpson 2012). As

Harrison-Pepper (1990: 127, 131) shows in discussing Washington-Square Park in New York, street performers intervene in the spatiotemporal organization of a space, with the "Performance [being] a dynamic, shifting, breathing event" which affects that space in terms of "density, accretion, durations, dispersal, and flow".

The outcome of such interventions is by no means certain. It might be the case that street performance offers something positive to these spaces in presenting an "urban ritual that challenges the way we think about public space by promoting spontaneous, democratic, intimate encounters" in some of the city's most "routinized and alienating environments" (Tanenbaum 1995: 1–2). This might mean that such performances allow "for a certain kind of sociality that comes from particular forms of gathering in public spaces" (Amin 2006: 1019). That might sound quite grand but it can happen in quite small and fleeting ways. It might start from something as simple as a small child starting to dance to the music being played and spread through the smiles of passersby who witness this response and stop a while to watch this scene unfold (see Simpson 2011a). Equally though, it might be the case that street performances do something else. Given the capacity of sound to spread significantly beyond the site of its generation and permeate space, and given that streets (and their surrounds) are occupied by a host of bodies with differing dispositions and agendas, such performances have the capacity to generate negative affects in relation with some of their (less willing) audiences who might assess them to be a noisy nuisance rather than some kind of enlivening event (see Simpson 2017c).

In addition to these capacities to shape the character of the spaces in which performances take place (for better or worse), the street also presents a very specific environment in which to both perform and listen. This applies to all forms of musical performance which happen across a range of contexts (see Morton 2005) but again this is amplified in certain ways when it comes to the street (see Simpson 2013). For example, while performances in a more traditional venue take place to certain schedules, in the street, the performance unfolds against the backdrop of the everyday rhythms of that space. This can impact upon the experience of performing. While regulations might timetable performances into defined slots, there are more subtle and unfolding rhythms here – the relative speed at which people walk, how crowded or quiet the street space is, the extent to which people linger or continue to move, and so on all can impact upon the performer and their performance. Also, the specific materiality of the performance space might matter to the embodied experience of performing. There is, for example, less in the way of demarcation between performance space and street space here, and so different sorts of interaction with the audience might unfold in this setting. Furthermore, the literal atmosphere in performing outside and in exposed spaces can matter – wind and drizzle might chill and stiffen fingers making guitar strings painful to depress and movements hard to achieve; sun, while initially enlivening, might come to burn skin and dehydrate bodies, again reducing capacities to perform.

In these senses, we see clearly the complex and unfolding interrelation between performances and events that can contribute to the production of specific space-times of experience and a broader context which can play a part in, though not necessarily dictate, the unfolding of such production.

Geographies of listening bodies

The practicing of music (or musicking) is potentially as much about the audience that listens as it is about the performers who produce the music itself. The audience too is a fundamental part of the musical event. This audience might take part in a live musical event such as the street-based performances discussed above, or a more traditional concert event, at a music festival (Duffy and Waitt 2011), or something semi-formal/ in-between like a pub-based music session (Morton 2005). Equally, that audiencing might take place in relation to recorded music listened to in other more frequently encountered settings like the home (Anderson 2004, 2005 and Chapter 3), while driving/passengering in a car (Waitt et al. 2017), or when walking down a street listening to music on headphones (Bull 2000). Whatever the specific setting and whether live or recorded, listening bodies can become wrapped up in a listening event which shapes the subject that listens, the capacities of the body that listens, and the character of the space where listening happens in potentially unpredictable ways (see Chapter 3). And this also extends to non-musical sounds which compose a significant element of our routine inhabitation of a host of difference spaces (see Duffy and Waitt 2013; Gallagher 2016; Revill 2016). What we can begin to see is the

> importance of the specificity of context, the active power of sound/ music and the ways that sound/music is integral to how space is enacted in the flows and encounters between bodies and affects/ emotions.
>
> (Waitt et al. 2017: 329)

When it comes to thinking about listening, this again has been in the background of geographic scholarship and has only recently emerged as a more sustained concern (see Doughty et al. 2016; Pavia 2018). Music has obviously been listened to by geographers when it comes to studying music, but this has been masked by a focus on other themes. Again, the focus for a long time was on what the content of that music could tell us about identity, relations to place, and so on rather than actual practices of listening. There are exceptions to this. For example, Smith (2000) discusses listening as 'passive reception' but also as an act of decipherment that partially gives meaning to the music. Further, Revill (2000b) has discussed the ways in which the sonic properties of music inform moral geographies of landscape, nations, and citizenship. This then moves toward the sound itself as well as more obviously textual features of the music. Connecting this work together, though, is a cultural politics of interpretation based on acts of conscious thought and deliberate reflection/decipherment that construct an understanding of its object (Schusterman 2000).

One way we can draw out the difference NRTs have made here to thinking about listening comes from the work of Jean-Luc Nancy. Drawing out a distinction made in French between *ententre* and *ecouter*, Nancy (2007) draws attention to a difference between 'hearing' and 'listening'. Hearing (*entendre*) connotes understanding or comprehension while listening (*ecouter*) does not have such connotations. This suggests that listening need not be tied to interpretation. Following this, work developing around NRTs has sought to decenter the role of interpretation in accounts of listening. That does not devalue the work that has already been done on 'hearing'. Rather, it suggests that there are also other important approaches to be taken. Such an approach to listening would "certainly not mean that we would abandon meaning, signification, and interpretation", but rather seeks to develop an understanding of listening that would allow us "to relate to the world in a way that is more complex than interpretation alone" (Gumbrecht 2004: 54). A key part of that complication is a concern for the affective capacities of sound, with sounds' expressive materiality (its resonance, timbre, rhythmic pulses, and so on (see Simpson 2009)) which can have quite 'visceral' impacts on bodies and the setting in which listening happens (see Gallagher 2016; Waitt *et al.* 2017). And to return to something of a refrain when it comes to such moves, this is not to remove political concern when it comes to practices such as listening (see Box 6.3). Rather, the concern is to approach listening in terms of an affective politics attuned to variations in "our capacity to listen and to respond to one another" which might emerge as much through attending to how the message is delivered as what is heard (Kanngieser 2012: 337; also see Brigstocke and Noorani 2016).

Where this work leads, then, is toward an 'expanded' form of listening which seems to "grasp the multiple effects of sound in the spatio-temporality of social life" (Pavia 2018: 82). Such an expanded listening attends to "the varied ways in which bodies of all kinds – human

BOX 6.3 DIFFERENCE AND THE EMBODIED EXPERIENCES OF TRANCE

Focusing on the experience of sound and music in terms of embodied experiences and affective relations raises questions when it comes to the importance of bodily difference and identity (as well as 'politics' more generally). Much of the work on music drawing on such ideas has tended to position itself in contrast to prior work in geography on music where the construction and reproduction of different identities was key. However, work here has also sought to rethink difference and identity in light of such ideas when it comes to the experience of music. A prominent example of this can be found in the work of Arun Saldanha on the rave tourism scene in Goa (see Saldanha 2005, 2007; also see Chapter 4). In this, Saldanha (2005: 707) starts from the premise that "corporeal difference needs to frame the analysis of music and space". To explore this, Saldanha focuses on a musical context where a host of different bodies come together to experience the same music and, in this, how a range of differentiations occur between these listening, dancing bodies.

Saldanha (2005) focuses on what he calls the 'morning phase', a period from dawn until the music ends. During this time:

> when it starts getting lighter in the east, the social forces of the morning phase becomes literally apparent as the Indians leave the dancefloor and the white tourists gradually take over the party. By the time the first rays of the sun pierce through the coconut trees, the dancefloor has become almost entirely white.

> (Saldanha 2005: 710)

In thinking about the processes of differentiation that race emerges through here, Saldanha (2005) focuses on a range of themes that emerge from his encounter with this scene. We hear of clashes of cultural expectation unfolding on the dance floor between domestic tourist ('drunk Indians') who inhabit a world of domestic alcohol prohibition and the white women that they grope. We hear of the transformations that bodies and psyches undergo as the sun comes up, the temperature and humidity changing, the music played shifts, and the dancing bodies become more clearly visible to those involved (and differentiated from each other as a result). One part of that visibility that Saldanha picks up on is the extent to which

(Continued)

the bodies experiencing this music are tanned. Saldanha (2005: 714) suggests that:

> An indispensable bodily marker in any beach scene is the tan. Morning phase discloses who has been in Goa longest, and who is therefore the coolest. This coolness is only available to white bodies. ... When white men take off their shirts during morning phase, or white women wear a bikini top, it is to maintain their tan, to connect skin to sun, to them a completely legitimate way to be (un)dressed at a musical event in the tropics. Domestic tourists are not used to doing such things; heat and open space are for them not associated with exposing flesh. And so they seek the shade or their bed, leaving the sunlit dancefloor to the beautiful bodies of the Goa freaks.

Saldanha connects this difference in relation to sunshine, heat, and (dis)robing to ideas of 'phenotype'. The phenotype of an organism relates to the observable physical characteristics and behaviors of that organism – its physical form or structure. Saldanha suggests that this is more than simply biological and rather is something that is a complex and emerging product of technologies (i.e. clothing), cultural stereotypes, and social relations tied to a specific place and time.

There are other ways in which bodies are differentiated here which are less obviously visible and rather occur at a more molecular level of materiality and affect (see Chapter 4). Important here is the use of different substances. Bodies become differentiated along the lines of which bodies use which substances. Domestic tourists drink alcohol (again, 'drunk Indians') and take cocaine. White 'Goa freaks' trip on acid/LSD. From this, Saldanha (2005: 715–716) suggests that:

> Goa freaks connect very differently to drugs, music, the landscape, celestial bodies, Indian culture and Indian poverty than domestic tourists or locals do. For freaks, Anjuna is psychedelic paradise, a place for continuous partying and chilling-out. They are clear about why they are there: to transform themselves, at least for some time, into gorgeous party animals and beach bums. They are so serious about this project of self-transformation that people who are not that committed to psy-trance – especially 'drunk Indians' ... – are looked down on.

In all of these relations, the key recurring point is that bodies here express in a range of ways and that such expression differentiates bodies. They express social relationships both through connections made and through the prejudices that they base (unspoken) differentiations upon. Some bodies become authentic and others not. Different bodies position themselves differently both in space and time. And these bodies express race through the aggregations that emerge in these encounters amongst and between certain bodies. This all unfolds through the encounter with this music scene.

and more-than-human – respond to sound" (Gallagher *et al.* 2017: 619). The concern here is to move:

> outwards from the dominant anthropocentric understanding of listening, beginning by deepening and expanding human listening (in relation to landscape), then considering how sound moves bodies beyond cochlear listening and human consciousness (as affects and atmospheres), and finally exploring forms of listening in which human bodies are marginal (vibrations in earth materials and machines).
>
> (Gallagher *et al.* 2017: 618)

Where this ultimately leads is an understanding of listening whereby concern is directed toward the responsiveness of that which encounters sounds, be that a human or animal body (not just a human or animal ear) or other forms of materiality which might be affected by sound and in turn affect sound's (re)sounding (see Ash 2015b).

Situating power in performances

These accounts of dance and music/listening have caused a stir in terms of their relation to a host of prominent themes and concerns in recent work in cultural geography. This has been most prominent in responses to work on dance practices (though see Revill 2004). Again, the question of what counts as politics looms large. It has been argued by Nash (2000) in the context of work on dance that, for example:

> Thrift ran the risk of paying insufficient attention to the social, cultural and spatial contexts within which specific dance practices were practiced; he therefore abdicated a thorough engagement with the politics of dancing bodies.
>
> (McCormack 2008b: 1825)

The emphasis found in such work on the present moment and happening of performance has been argued to "evacuate these larger sociohistorical processes of their political force and meaning" (Mitchell and Elwood 2012: 792). Such critical commentaries have been articulated by a range of geographers. For example, in discussing the efforts of urban reformers in the United States to provide recreational facilities for urban children, particularly in terms of the methods employed that aimed at identity formation through physical education (including dance), Gagen (2004: 420) was reluctant to draw on NRTs in thinking about "the practice of playgrounds" as she did "not wish to present the theories of playground reformers and physical educationalists as ontological truths". Instead, Gagen (2004: 420) chose to focus on "a historical moment in which children's bodies were reconceptualized in relation to consciousness" and from that examined "the spaces produced by this theorizing". Gagen (2004) sought to 'historicize' the moving bodies encountered in her research but do so in a way that focuses on those bodies themselves (in that historical context) rather than the symbolism that might be imposed upon them. In this sense, Gagen's (2004: 423) concern with NRTs is that "by celebrating the experiential presence of movement and physicality, they [NRTs] ignore the manifold ways in which the body's presumed immediacy has been used as a means to an end"; bodies are worked upon here in quite literal terms through the education of their physicality.

There is an interesting point of connection and contrast that can be drawn here between Gagen's concerns and subsequent work emerging on dance associated with NRTs. As has been suggested throughout this book, NRTs have evolved over time and have responded to various critical debates that have emerged in response to them. Like the ideas NRTs espouse, they themselves are dynamic. It's is clear, for example, in McCormack's (2005) account of dance that concerns over the training of dancing/moving bodies are something that can fall within the bounds of NRTs. Here, McCormack directly takes up the question of power in NRTs and asks whether this can be understood in ways that extend beyond certain accounts of, for example, performativity (see Chapter 1). This asks whether it might be possible for relations of power to unfold beneath the threshold of representational thinking but that nonetheless produce some kind of affect or effect in the world. In approaching this, McCormack (2005) explores Deleuze's conception of 'the diagram' which he sees as offering a means of thinking about how various affective forces and relations find consistency in the unfolding of practices in ways that both limit and enhance various bodies' capacities to act. McCormack (2005) explores this in the context of the practice of 'eurhythmics' – a set of principles developed with the aim of improving musical education through training bodily movements – which is seen to be a particular means through which the affective potentials of bodies come to be organized. McCormack (2005: 127) suggests that:

> The articulation of these principles can be understood as an effort to give consistency to a kind of kinaesthetic diagram. The effect of this articulation was not to produce idealised representations of the body. Nor, indeed was it to produce a form of alienated corporeality divorced from the actuality of lived bodily experience. Rather, these principles are best understood as an effort to diagram movement in such a way as to hold together, or give consistency to, the nonrepresentational powers of affective corporeality.

The point here was not to discipline bodies (in a restrictive sense) but rather to 'awaken' bodies through the development of an increased awareness of the rhythms of their movements. This is clearly then about relations of power when it comes to the conduct and actions of bodies, but that power is not something wielded over the body but is something that comes into relation with that body's capacities (either positively or negatively) (also see Simpson 2008). Such an engagement with the practices of bodies, thought in diagrammatic terms, marks out a field of potential in which we cannot know in advance how bodies will react; there is every chance that possibilities and capacities for action are opened up as well as shut down. Or there may be little affect at all. In this, "It is precisely this uncertainty that makes the moment political ... insofar as it is productive of a difference whose sense is not circumscribed by the territories of identity" (McCormack 2005: 143–144).

Moving to another prominent critique of this work on dance, Cresswell (2006a) is critical of NRTs' argued lack of attention to representations in work on dance and suggests their ability to produce 'correct' and 'appropriate' movement, and more broadly 'codify' and 'regulate' dancing bodies. Cresswell's (2006a) interests lie in the 'normative geographies' at play in the training of bodies, specifically in the context of ballroom dancing. Referring to McCormack's (2003) work on dance, Cresswell (2006a: 72–73) argues that "Although codified rules are important to McCormack, they are not as important as the styles and modes of moving themselves" and that "This notion lies at the heart of contemporary geographical discussion of non-representational theory and the idea of practice". By way of response, Cresswell (2006a: 73 [emphasis in original]) suggests a need to pay attention to the "interface between the representational and the non-representational", or what Dewsbury (2011a) describes as the 'historical context' of dance and its 'performative show', so as to uncover the ways in which "representation is used to hijack the process of *becoming*" (also see Revill 2004).

Leaving aside the repeated statements of those developing NRTs that such work is interested in representations in terms of what they do in the unfolding of practice (see Dewsbury *et al.* 2002), this is an interesting point which highlights a key point about how NRTs attend to representations. Echoing the discussion of McCormack's (2005) work on eurhythmics above, there is a different sense of the power of representations here and what they might do. NRTs do not pursue the sort of normative geographies Cresswell is interested in. This comes down to both the potential efficacy of representations to delimit practice – whether they can actually 'hijack' becoming – but also the idea that their role will be negative or constraining. Perhaps, we can reframe this critique in light of such arguments in a way that has featured in a host of work around NRTs for some time, meaning the question could become: what roles do representations play in the becoming of the world – both enabling and constraining – when it comes to the situated nature of practices within a specific context? Following Thrift (1996: 3), this context definitely does

> not mean an impassive backdrop to situated human activity. Rather …
> [it is] … a necessary constitutive element of interaction, something
> active, differentially extensive and able to problematise and work on
> the bounds of subjectivity.

Collectively, in these critiques we can see a "cautious skepticism about the ability of non-representational theories to grasp the politics of moving bodies" (McCormack 2008b: 1825). However, again, we dance around the question of what is meant by politics here and what qualifies as political. The position consistently returned to in various NRTs is that such politics are not given in advance and that the attentiveness to the affective, unfolding nature of embodied practices found in such NRTs might recognize a different politics.

Conclusion

This chapter has introduced work influenced by NRTs' concerns with what have been called 'arts of the now'. While work around NRTs has been interested in the ongoing composition of everyday life through the unfolding of a host of mundane, everyday practices, and at times has considered these 'as' performances, NRTs have meant that geographers have become increasingly interested in, and have taken inspiration from, actual artistic performances. This meant that artistic practices were some of the earliest objects of empirical experimentation for NRTs, but also that more conceptual inspiration has been found from both these practices and the scholarship in performance and dance studies that has emerged around them.

In thinking about performances as arts of the now, a range of key themes emerged in this revolving around the generative relations that take place in them between bodies and spaces. It has been shown here how dance and music can act to shape the experience of spaces, if fleetingly. It has also been shown how spaces can act upon the unfolding of such practices. Either way, the attention here shifts away from the 'content' of these performances – the meanings that might be found within that which is performed – toward the expressiveness of the performance practice itself. Meanings remain important to that unfolding expressiveness. However, performances' content does not exhaust their potential to do something in the unfolding of social life. Performances hold the potential to impact on the unfolding of social life in ways that exceed the meanings we might associate with their content (the meaning of lyrics, for example) and the social–historical context within which that performance practice developed.

These developments noted, this chapter has shown how these accounts of artistic performances informed by NRTs have come under criticism on a number of fronts. There has been a suggestion that focusing on these as 'arts of the now' risks forgetting the 'then', that we lose a sense of the historicized context within which performances unfold, and that such accounts are 'asocial'. Similarly, there have been calls to attend to the relationships between 'the representational' and 'the non-representational', specifically in terms of how the former might limit the latter. As has been noted several times in the chapters that precede this one, it is arguably the case that work informed by NRTs both does not neglect such points but rather considers them in ways which differ from prominent approaches in human geography/the social sciences more broadly. We again encounter the point that NRTs approach a different sort of politics. It is important also to note that being critical of how 'NRT' (in the singular) approaches any given topic can prove risky in the sense that it is a critique levied at something that is both loosely defined and itself still emerging and unfolding. In that sense, questions of the situated nature of embodied practices and performances have become an established, ongoing concern for geographers continuing the development of NRTs. It is likely that such concerns will remain established in thinking about the unfolding of practices and performances of various type/genre in the settings in which they take and make place.

Further reading

Rogers (2012a) provides a detailed introduction to work in geography on the performing arts as well as work from the performing arts that variously approaches geographical themes. This focuses on themes around landscape/ecology, place/site, and cities. The discussion extends beyond the bounds of work explicitly aligned with NRTs, but relevant work is discussed and situated in broader trajectories of scholarship on performance. **Revill (2004)** provides a critical reflection on studying dance in light of NRTs. This is based around auto-ethnographic research involving learning French folk dance and is illustrated with a series of reflections from this practice. The discussion here revolves around the tension/relation between 'the representational' and the 'non-representational' when it comes to NRTs and suggests that the focus should be on their intersections. **Wood (2012)** provides both an accessible introduction to a host of work from geography and beyond on music and makes clear how NRTs depart from such work. Further, this paper shows where these ideas lead by discussing actual (Scottish) musical performances. This emerges through an application of the ideas introduced in this chapter around 'musicking' in the context of two music festivals in Scotland, considering music from the perspectives of audiences, performers, and festival organizers. **Pavia (2018)** provides a review of scholarship on sound in geography, including recent work emerging around NRTs. This shows clearly the shifts that have taken place in terms of how sound (including music) has been studied in geography. Further, the review identifies work from a range of national contexts, including non-English-language work.

7

NON-REPRESENTATIONAL
THEORIES AND METHOD

Introduction

I want to return to the scene I opened this book with one more time as we move toward a conclusion. Rather than simply observe and reflect from an office window, though, we can reflect in a little more detail on how those working in the wake of NRTs might actually research such a scene. Based on existing trends when it comes to methodologies in human geography emerging in light of NRTs, we might encounter something like the following...

- *Methodology 1: auto-ethnographies of student life.* A researcher walks down the street on the way to the University campus. Standing at the crossing, waiting for the lights to change, they notice a sheen of sweat on the brow of the person stood next to them. At this point, they become aware of the humidity of the atmosphere and the limited breathability of their coat's fabric. This brings back memories of their 'fieldwork' in a night-club the night before – of being amongst moving, sweating bodies amid the cacophony of throbbing beats. These ruminations are interrupted by the arrival of two others chatting animatedly about what one of their friends did on their night out the previous night. Another stands at the crossing checking social media on their phone and typing a response with surprising agility on the tiny screen-keyboard. The lights change, the traffic stops, they move off and cross the road, realizing as they cross how little thought such habitual rule following takes. The researcher heads off to write down some of this as best can be remembered, knowing what was going on that night before and that morning will never be captured by the words in their research diary.

- *Methodology 2: videographic observation.* Another researcher stands someway off from the crossing behind a video camera tripod. They are checking emails on their phone, replying to messages, and working out when they will have time to analyze the hours of video material they are producing as the camera runs constantly alongside them. Later, they are seated in front of a computer screen reviewing the video recorded. They manipulate the playback speed, condensing the hours of footage into minutes and so starting to see the rhythms which characterize the everyday life of this space. They also watch, and re-watch, short sections of video, examining in detail the micro-scale interactions taking place between people moving through the space – smiles, waves, averted gazes, and so on – that might give the space a certain feel of conviviality or indifference.

- *Methodology 3: walking, talking, recording.* A final researcher walks toward the crossing with a member of staff from the University. They're coming to the end of their journey that started out at that person's home. Along the way they have talked about their life in this place and various features of the local environment (anything from architecture, to infrastructure, to social spaces, to outwardly banal but nonetheless significant places, amongst other things). The conversation has been recorded via a GoPro camera and the route recorded by a GPS app. Later, the researcher will work through the combination of comments, route map, and visual record while wondering how these materials might be meaningfully combined in producing an account of this route and these environmental relations.

Accounts such as these rarely get beyond caricature. However, each of the brief discussions above do reflect recent trends in the actual methods used in response to the challenges NRTs pose to doing geographic research. As Paterson (2009: 779) notes, NRTs do "not proffer any single methodology" and instead propagate "a raft of creative approaches and techniques, rethinking the usefulness and purpose of empirical investigations". Ethnographic and auto-ethnographic research has been a common response when it comes to trying to research things like affects and embodied practices (for example, see Bissell 2008; Emmerson 2017; McCormack 2002; Simpson 2013; Wylie 2002a). Video (amongst other recording technologies) has been used in a wide variety of ways in researching practices in a host of settings (compare Ash 2010a; Duffy and Waitt 2011; Duffy *et al.* 2016; Gallagher 2015; Gallagher and Prior 2014; Garrett 2010; Laurier and Philo 2006a, 2006b; Lorimer 2010; Simpson 2011b, 2012; Spinney 2009). And more established geographic methods like interviews have been animated both through the use of such technologies and in their transposition into the actual contexts they seek to discuss, as well as through creative attempts to present the material generated through them (Macpherson 2008; Latham 2003a, 2003b; Lorimer and Lund 2008).

These are by no means the only responses. However, I would suggest that they share common themes which both recognize and seek to respond to the challenges presented by NRTs. First, we can see attention to the actions and interactions of bodies in action. Either the researcher's own body and/ or those of 'research subjects' come under scrutiny as they actually do things in the moment of that doing. Second, we can see attempts to attend to the everyday activities and practices through which space is produced and which act to reproduce the day-to-day reality of its social life. That comes at anything from the level of micro-scale gestures, to the habitual acts of rule-following that bodies undertake, to the broader patterns and routines of activities that unfold in a western capitalist society. Third, each of these methods recognizes that all of the above is unfolding and therefore not straightforward to capture. Whether it be through the use of technology, situated-talk, embodied particip- ation, and/or creative presentation, there is a recognition that there is a lot going on here and that innovative attempts to attend to (if not entirely capture) that happening are a reality of doing such research. In that, each methodology not only employs different methods but they also very likely will offer entirely different accounts of what is happening in this scene.

In this chapter, I am going to explore such attempts to attend to the methodological challenges NRTs present to geographic research. In doing so, this chapter will focus less on the specific methods or recording techniques that might be employed by those engaged with NRTs. Rather, the focus will be on the broader methodological disposition that NRTs demand. The title of this chapter is perhaps misleading in this sense. This chapter doesn't give a 'how to' style guide to doing non-representational research; I won't get into which camera lens to use, which GPS app might prove most fruitful in recording movements, or what protocols an auto-ethnographic observation should proceed on. Rather, the concern will be with identifying aspects of what might be seen as a certain *style* of working that NRTs push research toward (Ash and Simpson 2019; Vannini 2015b). Before concluding, though, the chapter will also turn the question of writing and research 'output' in the wake of NRTs and the methodological styles they suggest.

Expanded epistemologies; changing methodologies

The expanded onto-epistemological realm opened up by NRTs has led to a range of questions around how scholars might 'do' research after NRTs. As Greenhough (2010: 41) notes, the

> recognition of a lively material world, which we come to know
> through active experience rather than passive observation, entails

a new way of doing geography. Rather than making, describing or mapping the world it now involves paying attention to, and engaging with, the ways in which dynamic and changing worlds are lived with and performed through the interactions of living and lively beings.

NRTs have brought a whole range of new concerns into geography's view. However, being 'in view' is part of the challenge here in that these concerns – process, affect, the pre-reflective, more-than human materialities, atmospheres, and so on – often exceed human abilities to fully perceive and/or articulate them. We are left with a lot of uncertainty and a recognition of the partiality of our knowledge. Emerging within a discipline that had previously been heavily indebted to representationalist modes of thinking that focused on the deconstruction of representations, discourses, and iconography, and where various verbal methodological approaches had come to dominate its methodologies (Crang 2003), this situation has left geographers without a clear direction when it comes to moving into, through, and back from 'the field' (assuming that is something that can still be easily delimited (see Dewsbury and Naylor 2002)). Or as Paterson (2009: 780 [citing McCormack]) puts it, "the difficulty lies in capturing the intensities of what is 'often below the cognitive threshold of representational awareness that defines what is admitted into serious research'" when our methods are primarily representational in basis.

An initial challenge when it comes to 'doing' research in light of NRTs is that the pace of theoretical development that has led to geography's onto-epistemological expansion has not necessarily been matched by expanded methodological reflection. Much of the initial work developing NRTs focused on theoretical development and spent less time thinking about how that development both opens up and necessitates new 'methodological horizons' (Lorimer 2010; also see Latham 2003a). As Crang (2003: 494) noted in a review of qualitative methods in 2003, "methods often derided for being somehow soft and 'touchy-feely' have in fact been rather limited in touching and feeling". Geographers were both relatively late in turning to the study of 'the body' and even later when it came to thinking through qualitative research practice after that eventual turn (Crang 2003). This situation led Latham (2003a: 1998) to suggest in one of the first explicit reflections on method to emerge in light of NRTs that: "[w]e simply to not have the methodological resources and skills to undertake research that takes the sensuous, embodied, creativeness of social practice seriously", something that was further hampered by geography's "unwillingness to experiment with techniques that go beyond the now canonical cultural methods: in-depth interviews, focus groups, participant observation of some form or other".

Responses to these concerns have emerged in recent years and can be seen, for example, across a number of qualitative methods progress reports published in the journal 'Progress in Human Geography'. For example, Davies and Dwyer (2007) note that while 'traditional' methods continue to be used by human geographers, changes were occurring in how such qualitative methods were being conceived and carried out. This meant that:

> In place of the pursuit of certainty in generating representations of the world, there is recognition that the world is so textured as to exceed our capacity to understand it, and thus to accede that social science methodologies and forms of knowing will be characterized as much by openness, reflexivity and recursivity as by categorization, conclusion and closure.
>
> (Davies and Dwyer 2007: 258)

Further, roughly a decade later, Dowling *et al.* (2018: 780) identify the further development of such agendas in terms of increased "attention to the non-visible, the non-verbal and the non-obvious", meaning a turn toward "methods and methodologies that enable researchers to grasp and grapple with that which has, at times, been invisible, denigrated and unimaginable".

Such concern with method after NRTs can be seen to have reached critical mass with the publication of a book-length reflection on 'Nonrepresentational Methodologies' (see Vannini 2015a). Striking for some will be the lack of discussion of specific research methods in the sense that we might expect from other books about, for example, qualitative research methods. This is explained by Vannini (2015b) through a distinction between 'method' and 'methodology'. 'Method' is taken to refer to the "tools through which we get data" (interviews, focus groups, ethnography, etc. commonly discussed in 'how to' methods textbooks) and 'methodology' is taken to refer to the broader research strategy for how that data is dealt with, alongside 'big picture' questions of epistemology that orientate the research processes as a whole (Vannini 2015b: 10). Vannini suggests that there is no 'non-representational method' but that there might be 'non-representational methodologies'. And what holds such methodologies together is a certain 'style' of working which is concerned with "territorializing, de-territorializing, reterritorializing, and animating life" and so "with issuing forth novel reverberations" (Vannini 2015b: 12). In working with events, relations, 'doings', affective resonances, backgrounds, and the like, this style is orientated toward the future, with experimentation, with generation, with animating discussions rather than fixing them in place (Vannini 2015b).

In terms of more specific examples of what geographers have done in developing such styles of working, there are a range of examples where such challenges have been taken up and where there have been attempts to inject more creativity and playfulness into social science traditions (Thrift 2003) (see Box 7.1). What we tend to find here is an attempt to "cultivate an affinity

---

BOX 7.1 NON-REPRESENTATIONAL ETHNOGRAPHY

Ethnographic research has become something of a hallmark of empirical work done following NRTs (for example, see Emmerson 2017; McCormack 2002, 2003; Morton 2005; Revill 2004; Rose 2010b; Saldanha 2005; Simpson 2011b, 2013; Wilson 2011; Wylie 2002a, 2005). There are aspects of this which show the sorts of attempts at creativity and experimentation that Latham (2003a) and Thrift (2003) call for. This can be seen in Vannini's (2015c) outline of what he calls an evolving 'non-representational ethnographic style'. Vannini's account here is less a 'how to' and more a reflection on some of the emerging characteristics of the ethnographic work NRTs have inspired. Rather than report back 'from the field' their "faithful rendition of the world 'as is'":

> non-representational ethnographers consider their work to be impressionistic and inevitably creative, and although they are inspired by their lived experiences in the field, they do not claim to be able, or even interested, in reporting on those in an impersonal, neutral, or reliable manner.

(Vannini 2015c: 318)

The agenda here in these non-representational ethnographies is to try to produce accounts of the world that 'make sense' based on the world encountered. This account, though, is produced in the recognition of the contingent, partial, and situated nature of that encounter and the inherent creativity that unfolds in that making of sense, in any encounter and any in accounting for it. Talking of partiality and contingency might sound like consigning any endeavor here to failure – that a 'true' account cannot be produced and so everything we do will fall short of actually accounting for the world – but that is not taken here as a negative. Rather, this is seen as a creative opportunity which should lead us to embrace experimentation and so to aim to try and 'fail better' (Vannini 2015c; also see Dewsbury 2010a).

In providing such 'animating' accounts, Vannini (2015c) suggests that it is possible to identity five interrelated and overlapping qualities of these ethnographic accounts of the world:

- *Vitality.* Rather than focusing on "rational behavior, politico-economic causation, cognitive planning, instrumental interaction, and mechanistic predictability" in their encounters with the lifeworlds of research subjects and trying to decode "textual data to give rise to explanatory descriptive categories", non-representational ethnographies are "pulled and pushed by a sense of wonder and awe with a world that is forever escaping, and yet seductively demanding, our comprehension" (Vannini 2015c: 320). The vitalist ideas that inform aspects of NRTs (see Chapter 1) emphasize the dynamic, and unpredictable, unfolding of life, and push toward a different approach to ethnography whereby researcher and researched, and the world they exist within, are seen to be (constantly) remade through the unfolding of the research.

- *Performativity.* Non-representational ethnographies emphasize

the importance of ritualized performances, habitual and non-habitual behaviors, play, and the various scripted and unscripted, uncertain, and unsuccessful doings of which everyday life is made, no matter how seemingly mundane or unimportant

(Vannini 2015c: 320–321)

This means that ethnographies are "not so much undertaken to explain but rather to make audible silences and invisible forces" in the talking place of situated actions (Vannini 2015c: 321). Such actions might be dramatic but performativity here isn't a synonym for performance; this is much more about the un-reflected upon everyday (re)making of social worlds as the conscious construction of them.

- *Corporeality.*

Non-representational ethnographic research begins from the researcher's body as the key instrument for knowing, sensing, feeling, and relating to others and self. Passions, orientations, moods, emotions, sentiments, sensations, dispositions, colors, sizes, shapes, and skills work as the bodily fluids enlivening all relations in which ethnographic relations are entangled. From fatigue to enthusiasm, melancholia to keenness, pain to enchantment, non-representational ethnographic research is affected by bodies' capacity to affect the world and their capacity to be affected by it.

(Vannini 2015c: 321)

This focus on affective relations is not just about the researcher and the researched bodies (be they human or non-human, animate or inanimate). The point is that non-representational ethnographies

(Continued)

are meant to be affective in terms of what they show or evoke about the world (see Stewart 2007).

- *Sensuality*. Vannini (2015c: 322) suggests that:

> Non-representational ethnographies underline the not-necessarily reflexive sensory dimensions of experience by paying attention to the perceptual dimensions of our actions and the habituated and routine nature of everyday existence.
>
> (Vannini 2015c: 322)

In so doing, "they engage in sensuous scholarship: research about the senses, through the senses, and for the senses" (Vannini 2015c: 322).

- *Mobility*. Non-representational ethnography means being attentive to movement – both actual and in terms of the dynamic unfolding of everyday life – and so situating "fieldwork in the concrete time-spaces that ethnographers actually inhabit" (Vannini 2015c: 323). With this, the ethnographer starts to 'wander' and

> These wanderings are also wonderings which seek out the interweaving storylines binding self, others, places, and times – lines which, just like ethnographic travel, are dynamic, unpredictable, with no clear roots or obvious boundaries or ends.
>
> (Vannini 2015c: 323)

for the analysis of events, practices, assemblages, affective atmospheres, and the backgrounds of everyday life against which relations unfold" (Vannini 2015c: 318). For example, in an early response to these challenges, Latham (2003a: 2000, 2012) sought to develop a "broadminded openness to methodological experimentation" by imbuing traditional research methodologies with "a sense of the creative, the practical, and being with practiceness" through his experimentation with diary-photograph and diary-interview methods. Precisely showing that the issue here is not about finding 'new' methods, Latham experimented with these diary-based methods through combining them with the space-time diagrams of Time-Geography (see Chapter 1). This led to novel, multi-media outputs which give an impression (rather than definite representation) of the lives of the subjects and spaces Latham was interested in, both as part of the reflection on the 'data' initially collected and in the presentation of such research materials.

Further, in another early example, Morton (2005: 663) developed an experimental and loosely structured multi-method 'performance ethnography' when researching the spaces produced in and through the performance of Irish folk music. The aim here was to try to get at the "non-verbal, expressive and emotive, non-cognitive aspects of social practice and performance" (see Chapter 6). This entailed a form of 'observant participation' in these music sessions composed through audio and video recordings of the music performed as well as photography in the performance spaces and surrounding city. Further, musicians completed spoken diaries and were informally interviewed. What set this apart, though, was the specific disposition toward the sessions being researched – the focus was on the sessions themselves as events and their making rather than on them as 'things' that might be accounted for – and the presentation of this material in evocative ways (though multi-media outputs).

Ultimately, initial methodological developments here have grappled with the questions posed by NRTs of 'traditional' social science research by experimenting with both what is done in 'the field' and what is made of this; with what data is recognized as being potentially relevant and what sort of claims can (and can't) be made from that; and, have re-framed the researcher's position in the field, both in terms of discourses over positionality and reflectivity, but also in terms of the researcher's body itself becoming "an instrument of research" (Crang 2003: 499; also see Paterson 2009). We increasingly find geographers engaged in the practice they are researching rather than just talking to those who participate in them, observing those practices, or examining the representation of them in various forms of media. Practices are encountered as unfolding events which play a part in the composition of space and time from within the midst of that (ultimately elusive) unfolding.

Moving-image methodologies

Video has been a prominent feature of the methodological responses made by geographers to the challenges presented by NRTs when it comes to studying practices (see Ash 2010; Laurier and Philo 2006a, 2006b; Morton 2005; Simpson 2011b; Spinney 2009). The emergence of this response initially coincided with the increasing availability of affordable, high-resolution video cameras and user-friendly editing software. However, this has advanced significantly in recent years making it increasingly possible to embed video-recording technologies into research practice more and more easily and affordably. This can be seen in the emergence of 'GoPro' cameras and other wearable technologies (see Vannini and Stewart 2017), for example, not to mention the significant advancements that have taken place in 'smartphone' cameras. This, in turn, has allowed video-based materials capturing quotidian events to circulate more and more widely, notably through social media sites like 'Vimeo' and 'Youtube' (see Laurier 2016).

Geography's engagement with video methods has been motivated by the argument that:

> video can help participants and researchers alike bring into focus aspects of practice that have previously been blurred or out of shot. The advantage of incorporating a medium like video into the existing methodological tool kit is that the researcher can begin to explore how people use space and their bodies, how people interact with space, understand where and how they look, and ultimately gain a far more nuanced idea of how participants derive meanings through movement.
>
> (Spinney 2009: 828)

Video arguably allows geographers to pay closer attention to the actual unfolding of events in being able to watch and re-watch them. In doing so, video allows geographers to get closer to that unfolding in gaining (close to) the perspective of the bodies engaged in those practices. This offers something beyond the ability to 'write' events in diaries or record commentaries on them through Dictaphones.

Significant when it comes to geographic engagements with video methods has been Eric Laurier's use of video in a host of empirical contexts. Laurier has sought to explore individuals' skillful negotiation of a variety of mundane everyday tasks and practical activities, ranging from finding a parking space, to talking on the phone, to going for a coffee (Laurier 2001, 2005; Laurier *et al.* 2006). In this, Laurier has often used 'naturalistic' forms of video recording. While recognizing that the act of filming may well intervene into that setting making the events encountered less than 'natural', this approach involves setting up a camera in a given space and leaving it running for a period of time (Laurier and Philo 2006c). From this, Laurier makes fine-grained, story-boarder accounts of how, for example, sociality happens in a café, focusing on banal features like individuals' body language, gestures, facial expressions, brief comments, and the like and how these play a part in shaping the conduct of everyday life in such spaces (Laurier and Philo 2006a). Starting from the insights of ethnomethodology and conversation analysis, and so being concerned with how it is that people (re)produce the social orders of everyday life, Laurier's work has illustrated the potential of video as a recording tool in the study of practice; it allows researchers to repeatedly view and reflect on the minutiae of apparently simple and mundane everyday activities – to get a sense of the practical knowledge present and through which life is 'done' – and so attend to things which might be missed when observing the actual happening of that practice in real time.

Further, and specifically engaging with themes around affect and more-than human relations, Lorimer (2010) discusses moving image methodologies – involving both the collection of video material and the analysis of existing video (i.e. films) – and their potential to grasp the more-than and non-representational dimensions of life (see Box 7.2). In particular, Lorimer (2010: 242) is interested in the ability of such media to witness, make sense of, and evoke human–nonhuman interactions which, he argues, "escape text- and talk-based approaches". In doing so, Lorimer focuses on a range of complex and uncertain bodily interactions between humans and non-humans (in this case, elephants) that can be viscerally 'sensed' and 'felt' through the watching of video clips and may 'echo' the gut instincts of the observer/camera person. Something that becomes apparent here is the necessity of engaging with video methods in an embodied way (see Paterson 2009), both in terms of the bodies being watched but also the body (or bodies) watching.

In addition to video recording as a means of attending to the unfolding of a range of events and practices, Lorimer's attention to the analysis of existing media material highlights another key point here – video material can be (increasingly and easily) manipulated. This obviously holds implications when it comes to working with 'secondary data' and, for example, thinking about how films and other media might generate certain types of affect and effect as well as represent 'realities' (see Carter and McCormack 2006; Simpson 2014b). However, such manipulation can also form part of the analysis process itself. Changing the playback speed, for example, can shift our point of view and bring to light a host of rhythms and patterns which at first (or second, or third) look, might not be so apparent when it comes to the composition of day-to-day living. This is not something 'new' when it comes to studying things like embodied practice or everyday life. Muybridge's photographs of trotting horses are a well-known example of this (see Cresswell 2006b), as is William H. Whyte's use of time-lapse photography in studying the workings of the 'small spaces' of cities (Whyte 1980). However, with the increasing ease of such forms of recording/editing, such manipulations might offer something when it comes to attending to the multiple durations upon which certain events or practice unfold (see Simpson 2012).

Do we need 'new' methods?

Sometime toward the end of my PhD – I think it was sometime in 2008 – I remember sitting in a conference session toward the end of a large international geography conference listening to the nth paper of the week that had included video as part of the methodology. Many of these papers had referred to developments around NRTs and associated concepts. More than the content of those papers – which were all well put together, original in

BOX 7.2 'MOVING' IMAGES

While moving images have formed a central part in geography's response to the methodological challenges posed by NRTs given the way that they help attend to the ever-unfolding nature of the world, that is not to say that still images have not featured in these responses, also (see Keating 2019b). As Latham and McCormack (2009) have shown, such images can 'do' a range of things beyond representing the world, especially when we start to think about 'photographing' as well as the photo itself. Taking a photograph – composing, framing, exposing, timing, etc. – is a creative act in and of itself rather than the production of a neutral representation, as is its ultimate production and showing. For example, in their reflections on a student fieldtrip, Latham and McCormack (2009: 252) found that such "images afford opportunities for attending to everyday ecologies of materials and things" through the ways that student generated (surprisingly consistent) inventories of the urban – photographs of door buzzer after door buzzer, bike stand after bike stand, and so on. These images obviously represent certain objects but when taken together, they do something else; they give insights into how the city is encountered in practice.

Such images also offer opportunities when it comes to producing an 'affective archive' through their creative presentation alongside other forms of written and visual media, extending beyond 'traditional' written outputs (Latham and McCormack 2009; also see Wylie 2006c). For example, Doel and Clarke's (2007) discussion of montage is illustrative. They see montage as "the process of selecting, editing, and piecing together separate sections of imagery for a calculated affect" (Doel and Clarke 2007: 890). Rather than (just) being about the 'content' of such images, this affect is generated through "the kinaesthetic jolts, estrangements, and disfigurements" that unfold when various images are juxtaposed alongside each other (Doel and Clarke 2007: 891). This is something encountered as part of our everyday lives and so forms something of our 'optical unconscious' that might become the subject of analysis, but montage might also be something deployed in and through research that could tell us something about the specific social conditions we live in by both producing and exceeding what we might see as "matters of fact" (Doel and Clarke 2007: 902).

focus, and very interesting to hear – the thing that I recall hearing again and again was a range of similar phrases following the playing of a short video clip of research participants doing something. The refrain of this conference seemed to be: 'So, as you can see…'. This was said each time in a way that made me question what I had(n't) seen. The thing that struck me here was thàt: (a) what was being talked about (affects, feelings, agencies, and so on) wasn't obviously visible in the clips (to me, at least) and (b) how much of a methodological norm including such video was becoming. We seemed to have reached a point where there was acceptance that video could capture such dimensions of practice, and so video had become a 'go to' method for a host of academics working in the wake of NRTs. The thing that troubled me was that there didn't appear to be a great deal of reflection here. In spite of various comments (noted above) that there is no such thing as a 'non-representational method' and pushes against seeing certain methods/ techniques as 'the answer', video seemed to be becoming (if tacitly) the agreed-upon answer, with the methodological sections of these papers variously noting something like: 'oh, and to get at these embodied aspects of research, in addition to interviews and participant observation, I used video'.

Video was something that I had, also relatively unthinkingly, turned to in my own research at the time in studying street performances (see Simpson 2011b). Videoing performances and the audiences of my performances seemed like something that was appropriate to do in thinking about how performances 'worked', how performers and audiences interacted with each other, and what sort of role the setting of the street played in these aspects of the performances. Certainly, such video helped me reflect back on my own auto-ethnographic experiences of performing; I could go back and re-watch encounters and reflect on what was going on, what it was about that specific encounter that had affected me so much (or so little), and so better understand the duration of the events playing out here. As I noted shortly after completing that research and attending the conference above, "video facilitated my reflection on the relations that occurred and what complex-singular relations contributed to their coming about, or was useful in the illustration and examination of my auto-ethnographic experiences" (Simpson 2011b: 350). It helped me in certain ways but it still wasn't 'the answer' to doing research after NRTs (also see Paterson and Glass 2020).

These critical points are not intended to suggest that the use of video in those projects did not bring something to either the research process or the ability to reflect in more nuanced ways on what had happened in that. Again, video evidently can offer geographers a range of things in their research practice. Rather, it draws into sharp relief what to me is the key issue when it comes to doing research after NRTs: that NRTs do not lead us to the use of specific methods or a single, defined methodology. Sticking with the use of video, while video has offered possibilities for many researchers, this video is "multisensory only in so far as they record video and audio" (Paterson 2009: 784). From this, then:

> while video can in some cases 'present' a certain affective experi-ence, 'bear witness to phenomena that often escape talk and text based methods' (Lorimer 2010: 251), or 'capture the "sparkle" and "character" of an event' (Rosenstein 2002: 6) ... in some cases video does not capture many aspects of practice that geographers have come to be so interested in. While in video we can see the minute detail of bodily movements and non-verbal communica-tion, alongside verbal communication and other sounds, it argua-bly provides little in the way of a sense of the felt aspects in and of these movements.
>
> (Simpson 2011b: 350)

Where do we go from here?

If video does not necessarily 'show' or 'present' the affective relations un-folding within the practices and events being researched in the wake of NRTs, does not provide 'the answer' to the challenges such ideas present, and instead more likely aids us in (incompletely) documenting and reflect-ing on the relations that occur in such events, where does that leave us? And further, what are we to do if "we never quite get at what is going on in the interviewee's head, nor are we ever so prepared not to be surprised by some revelatory event" and so are left "without a perfect representation at the end of it all" (Dewsbury 2010a: 321)?

Well, in the first instance, it leaves us with a lot of options as there is noth-ing inherently wrong with much of the methodological toolkit that geogra-phers have employed in a range of ways for some time. There is no reason why geographers cannot work with those methods. While some of the initial articulations of methodological options after NRTs talked about the limita-tions of talk-based methods, telling us, for example, that "interviews happen after the fact ...[and so]... can only ever provide an unsatisfactorily washed out account of what previously took place" (Hitchings 2012: 61), that is not to say that people can't talk about practice. Rather, we might learn from such (challenging) efforts to talk about the normally unthought. As Hitchings (2012: 66) suggests, geographers should

> reject the common qualitative interview at their peril. This is be-cause interviews offer such an efficient means of understanding

how it is to embody certain practices when it may be exactly such understandings that could prove crucial in initiating positive change.

While methods like video might get at something different than an interview, "A well conceived set of interview questions might well be far more effective at capturing the tension of the performing body as witnessed by the body of the interviewee" (Dewsbury 2010a: 325). Again, we find ourselves at the point of distinction between method and methodology. It is not that one method is 'better' than another, but rather that each method offers something different depending on the style with which it is deployed (on its own or in combination with other methods).

By way of something more instructive in finding ways forward, Dewsbury (2010a) offers a series of 'injunctions' which might guide the style of research we do in the wake of NRTs:

(1) don't fret about the risks of experimenting, it is a justifiable way of proceeding that works better if you really embrace it.

(Dewsbury 2010a: 321–322).

This starting point means that there are no prescriptions. Rather, Dewsbury's injunctions are meant to act as 'proscriptions' which broadly act against the 'scientism' under which social scientists find themselves placed; they push us away from a certain version of contextualizing, expecting, and following procedures which bind and blind us to certain forms of proceeding and knowledge generation without foreclosing where such a push might lead us.

(2) make sure you have conviction to stretch and strive to the full.

(Dewsbury 2010a: 323).

We need to avoid aiming to find 'the best' way of doing research in light of NRTs. It is not the case of developing a new form of methodological protocol that we (and others) will then follow or 'apply'. Instead, Dewsbury (2010a: 323) advocates that we imagine that we work in a lab or studio, testing out ways of proceeding, all the time embodying the "the ethos of stretching the means by which research is done and striving to continue as experiments fail or always fall short in the attempt".

(3) don't fear the judgement that tethers social science, especially that which is in close proximity to the humanities, to scientific values of efficacy and rigour.

(Dewsbury 2010a: 325).

We are all introduced in methods classes to themes around sample size, sampling strategy, representativeness, and so on. There are written (or unwritten) expectations over how much we should do, who we should include, how we should attend to differences in who we include, and so on. It is suggested that we need 'evidence' to back up our theories (Lorimer 2015). The same goes for things like what makes research ethical or not; there are very clear, almost unquestionable protocols to be obeyed. There is a lot of pressure to meet such expectations (from markers, funders, other academics, and people in 'the real world') meaning that it isn't necessarily easy or comfortable to push back against those in the pursuit of experimentation.

> (4) remember we are producing an understanding of the world because the world is not already out there as such.
>
> (Dewsbury 2010a: 326).

In carrying out research, we engage in a process of 'configuring worlds' through the ways that we frame our questions, the lens we take, the things that appear (and do not appear) to us in that research, what we report back, how we report it back, and so on. And making things more complex still, these worlds themselves are always in the process of coming into being; they are not fixed realities in space and time but unfolding ones, and this is something to which we need to attend. As such:

> methodology is ... about questioning how we are going to configure the world, and how we question in practice to what extent we are able to configure different worlds.
>
> (Dewsbury 2010a: 324)

> (5) diagram quite straightforwardly the space-time connections you experience with a palette and sensibility akin to the artistic.
>
> (Dewsbury 2010a: 328)

Less a proscription and more offered in the vein of 'why not?', the agenda here is not to produce a 'true account' of the world as it exists 'out there' but rather to find ways of engaging with the world that do something – that encourage us to think, that might impact upon how we see a particular space, practice, social setting, and so on. More data collection then doesn't mean a move toward a clearer, more coherent, more definitive picture of the situation; rather, such collection proliferates, opens out, multiplies implications and potentials for inducing thought.

(6) instead of concentrating on the cultural product, concentrate on
the cultural experience

(Dewsbury 2010a: 331)

Another way to put this would be to draw out a distinction between the
content of what is presented from our research – the evidence we have from
the field, be that interview transcript, video content, or whatever – and how
our research 'output' is expressed in ways that might seek to do something
more than report. What impact are we seeking to have on the audience of
this research? What are we trying to make them feel or do? What work are
we seeking them to do in their encounter with this research? Without deny-
ing the importance of representations and all that they do, the injunction for
methodologies after NRTs is to recognize that:

> we can as easily emphasize sensation over representation in the first
> instance, whereas it is perhaps the habitual status quo to start with
> representation which thus avoidably but unwittingly relegates the
> import of sensation itself.

(Dewsbury 2010a: 331)

Without offering a set of prescriptions here, it's clear that such a direction for
research practice after NRTs doesn't leave us in an 'anything goes' situation.
Rather, we are left in a situation where:

> the thinking, sensing, presenting aspects should be all specific
> interferences – interferences in problematizing how we think the
> world and how the world forces us to think, in attending intensely
> to the fluid, nervous, fleshy dispositions of our body's agency, and
> in how the world records itself on its surfaces both on the skin
> and in the cell, and in experimenting with the images we produce
> in disseminating our research across an open and mutually trans-
> forming nexus of expression, content, form and audience effect.

(Dewsbury 2010a: 332)

Writing non-representational research

Dewsbury (2010a: 332) ends his discussion of researching after NRTs with
a seventh injunction, to "remember you cannot directly signify that which
is past, so be more acute and cute in the research stories that you tell". This
raises the question of how research is presented in light of NRTs. In addition to such methodological challenges and the responses that have emerged

in light of them, one further challenge encountered when doing research in light of NRTs is the presentation of such research. How can we write about the world in a way that gives the impression of being "ensnared in the midst of life taking place, full of responsiveness and suggestiveness?" (Lorimer 2010: 185). The ideas here – and particularly their emphasis on the extra-textual, the fleeting and dynamic processes that lie beyond the reach of symbolic understanding, and so on – pose challenges when the traditional form of academic output usually takes the form of a written text (a journal article, a textbook chapter, a research monograph, and so on) which comes with certain expectations (a review of literature, a statement making clear its contribution to knowledge/that literature, (perhaps) a discussion of methodology, and the presentation of some sort of findings). Academic outputs remain limited for various reasons to such forms meaning, in an apparently contradictory sense, we still tend to only read about sensations, affects, atmospheres, events, and so on (Dowling *et al.* 2018).

Scholarship emerging around/after the emergence of NRTs has been concerned with questions of writing in a number of senses. There have been evident attempts to write differently in ways that are more overtly creative and experimental in style, form, and tone. Rather than attempting to communicate findings effectively (something not necessarily easy in itself), here, there is something more like a concern for the expressive potential of writing itself. We encounter "writing as word-strain" (Roberts 2019b: 1). In this, "At its best, non-rep writing creates a certain shimmer, as if altering the limits of perception, demanding that you burrow deeply into your own being and, paradoxically, making you lose track of yourself" (Lorimer 2015: 180). However, for some, such writing can present an elusive read that doesn't necessarily alter perceptions or force thought (see Cresswell 2013a).

In pulling apart this tension in writing in the wake of NRTs, the remainder of this section will focus on two trends in such attempts at research-writing, namely: performative writing and poetic/narrative accounts, and outputs that extend beyond textual means of presentation.

Performative and poetic/narrative writing

Performative writing has a complex genealogy with its origins in part coming from performance studies/the performing arts (see Phelan 1997). In broad terms, while encompassing a range of actual forms, a key principle of performative writing is that it seeks to be less a form of 'reporting' and more a form of performance; the point is to recognize that the act of writing itself, and what is ultimately written when encountered, can enact something. The reference to 'performativity' here recognizes that something (a word, a sentence, a turn of phrase) is about more than its meaning-based content;

it can also 'do' something that both repeats/reinforces some habitual sense of thing and plays with them (see Chapter 2). While 'scientific' styles of writing are (meant to be) about the clear contextualization of the research, the reporting of what data was collected, and the analysis of its meaning/ significance, performative writing strays from such attempts to represent reality 'out there'. Performative writing transparently strives to make written outputs active participants in our sense making (see Harrison 2000).

An early example of this emerging from NRTs can be seen in the opening of Thrift's (2000) article 'Afterwords'. Following a brief opening comment explaining that the article emerges in the wake of his father's death and the challenge of writing that event in a way that doesn't just make "him into fodder for yet more interpretation, by colonizing his traces", Thrift (2000: 213) speaks of a desire for "a form of writing that can disclose and value his legacy". Following this, we encounter a modest response: a short section titled 'Ghosts' which reflects on Phelan's (1993) account of 'the unmarked' – that which is no longer present but is still palpable in some way – and questions about what is visible/representable. Here, the text fades from black to gray. Much of the rest of the paper (baring one other fade-out) unfolds in a fairly typical manner for Thrift's articulations of what NRTs might be at this time: a wide range of references to social theory and philosophy; length direct quotations from such work; sections and sub-sections internally structured around numbered points of varying length; and, following the 'what is NRT' discussion, discussion of some kind of object/practice which leans to the thematic rather than what might more commonly be expected of an 'empirical' example or case study.

When I first read this piece (as an MSc student), I noticed the visual effect but didn't immediately recognize or make the connection to creative attempts at writing. Rather than an online PDF, I was reading a photocopy of the article made from a physical issue of the journal and, at first, I simply thought it was a product of that (flawed, grainy) reproduction. It was only later in the paper when I encountered the second fade that I was made to pause and go back. This is not meant as a critique of Thrift's modest attempt to play with writing here. However, it does draw attention to a key point when it comes to such performative writing. Performative writing might fail, might not work, or might work differently than its author intended. Again, the point here is not to foreclose things into a neatly presented package which reports a world 'out there' to the reader. Rather, what is written becomes something which is re-enacted when encountered by a reader, which may or may not produce thought, and may or may not lead to something beyond the content of the words on the page emerging.

Perhaps the most sustained attempt to develop such performative styles of writing in geography in light of NRTs can be seen in Derek McCormack's work (for example, see McCormack 2002; 2014; 2015c). Across a range of pieces, McCormack has developed a sort of impersonal style of writing where "the question of how to think with/in the world – this time, on this occasion, under these circumstances – is never settled in advance, but must be worked out, per-formed" (McCormack 2015c: 88). Through a series of experiments in style, McCormack focuses on various 'things' (rhythm, moving bodies, air, balloons, and so on) in an attempt to draw out the forces and relations that circulate around the circumstances they take place within. The style might be called 'multi-biographical' in that it mixes a concern for ethnographic encounters with such things but also "the diverse forces and participants that shape matters of collective interest as they register in various habits and bodies" (McCormack 2015c: 91). This 'circumstantial writing' moves quite far beyond the traditions of more scientific academic writing in that we encounter, for example, a series of short discussions, sometimes including theoretical reflection, that do not add up to a coherent whole (see McCormack 2014). Rather than holding theories and the empirical apart in their own sections or subsections, we start to see "the empirical as a field, or fields, of variation, with the important reminder that thinking is already and always a variation in this field" (McCormack 2015c: 94). In encountering this, impressions emerge, connections can be made, and a consistency to an idea emerges (or doesn't).

Extending such concerns for writing that goes beyond academic reporting toward something more creative that does as well as (or rather than) reports, a second emerging line here relates to poetic and narrative accounts of the world and the events unfolding therein. Geographers have increasingly been telling stories (see Lorimer 2006; Rose 2016). In this, geographers have started to cultivate a more "poetic sensibility… for evoking and describing sensuous dispositions and haptic knowledge" which "benefits from the styles and methods involved in experimental or creative writing" (Paterson 2009: 785). There are a number of points of reference and inspiration here. Perhaps, most prominent here become the writings of Kathleen Stewart whose ethnographic accounts of 'ordinary affects' tell stories of lives, encounters, and events that variously evoke, make strange, or potentially give consistency to a range of scenes from contemporary American life (Stewart 2007, 2010; see Fannin et al. 2010). That said, inspiration has also been drawn from other genres and authors, including literary figures like WG Sebald and Tim Robinson for the geographies of landscapes and spectrality they produce (see Wylie 2007b, 2012), or the 'Beat Generation' for their experiments in flowing, unedited 'spontaneous writing' (Honeybun-Arnolda 2019). In light of such work, we see a general trend toward expressive forms of writing which often rely on narrative forms of account when it comes to the events and encounters under consideration (see Box 7.3).

BOX 7.3 NARRATING LANDSCAPES

Wylie's accounts of landscape (see Chapter 5) present one of the most sustained efforts in work concerned with writing in the wake of NRTs, specifically in terms of developing a more literary, narrative style. Amongst these, we encounter language which departs quite substantially from what might be expected from a 'social science' style of writing. For example, in recounting a walk around 'Mullion Cove', we encounter the following opening words:

> The light ahead was so compelling that we were unstrapping our seatbelts, reaching for doorhandles – we were halfway out of the car before it even came to a stop, turned into a small gravel recess there by the cliff-edge. Sometimes you'll turn a corner and a view will surprise you, but we ran right up to this one, and then stood, together and apart; different angles on the same encircling scene.

(Wylie 2009: 289)

There are a number of features here that set this description of an encounter with landscape apart from what might be expected in a more traditionally framed account. First, the account starts in the midst of an encounter. We are not given a formal contextualization as we might expect – an outline of a study site or formal introduction of the 'issue' that the paper is addressing. Second, in this midst, the tone overtly sets the excitement of the scene, showing the affects and effects of the specific scene coming into view with impatient bodies. The prose here is not sober or neutral. Third, the 'we' here is ambiguous; the positionality of the researcher (and whoever they are with) isn't introduced; again, we're in the midst of an unfolding encounter. The whole point here seems to try to bring us back to something of the moment of encounter with which the paper is concerned. Fourth, this 'we' highlights an interesting tension in the context of NRTs in the way that it draws attention to the subjectivity of those encountering the landscape; in emerging in the midst of an arrival, it's clear that these bodies have come from elsewhere, have histories, and that these subjects are disposed in a certain sense to what they are arriving at. Again, we don't get a detailed enumeration of backgrounds. Here, the bodies-subjects and landscape appear to be coming into being through the unfolding of the encounter we get a glimpse of here. The ideas that lie behind the paper's subsequent unfolding are embedded in (and in turn shown by) the style and narrative being pursued (also see Lorimer and Wylie 2010; Rose 2010b; Simpson 2015b).

In addition to such a literary sensibility, this emerging style of writing after NRTs has at times involved the writing of actual poetry. For Boyd (2017a: 212), "Effective research poems are rooted in the sensual, have emotional poignancy, show a range of nuanced meanings, evoke empathic responses, and display an open spirit of imagination". Further, such geographic–poetic writing can start to experiment with the textual presentation of the poem itself and so embody something of the sensory concern the poem seeks to express. We can see this in Boyd's (2017a: 216) discussion of her research poems where the words follow a meandering path back and forth across the pages, "a path for the words that…bore some semblance to the aesthetic of the practice". Further, while by no means someone who would align themselves with NRTs and rather is an 'interested skeptic', Tim Cresswell has published a range of poetry which seeks to combine a geographical sensibility with poetic modes of writing and expression (see Cresswell 2013b).

The point here is to move away from providing 'empirical evidence' in the account given of a space, an encounter, or an event. Instead, through the presentation of "snippets of action, aura and atmosphere", the aim is to find ways of writing that "might recognize just a few of the ways that life takes place with affects in its midst; or, more radically speaking, how life is composed in the midst of affects" (Lorimer 2008: 552; also see Lorimer and Wylie 2010).

More-than textual accounts

While there are a range of creative attempts to experiment with what methods geographers use and what methodological styles they are deployed within, the outputs from such research do remain relatively standard: journal articles, book chapters, textbooks, and the like (Dowling et al. 2018). There are a host of things that restrict academics here beyond creative will or capacities. The contemporary audit culture of academic life leans on certain indicators of productivity and esteem. Research assessment exercises run by Governments involve the auditing of 'outputs' which, in social science disciplines at least, are commonly equated with peer-reviewed written texts. Promotion/tenure processes often focus on such matters in terms of volume and the 'quality' of such specific types of output (perhaps measured by proxies like citation counts over actual engagement with what is written). In other disciplines like the arts, more 'practice-based' or 'creative' outputs might be included. However, in geography (at the time of writing, at least), that is not normally the case. In spite of such institutional constraints, there have been signs of academics seeking to both move beyond solely text-based outputs and/or to produce other forms of research output.

Geographers working in the wake of NRTs have sought inspiration when it comes to methodology and the 'output' of their work from the performing arts and theatre practice (see Chapter 6). Here, a range of scholars have used theatre as part of their methodology, often within an ethnographic/participatory/collaborative research framework (see Raynor 2017a, 2017b, 2019; Rogers 2010, 2012a, 2012b). This includes theatre practices becoming an experimental and creative part of the research process itself, but also part of the output of research in terms of the actual performance pieces. As Raynor (2019: 692) explains in reflecting on her experiences with theatre-based collaborative research, in the unfolding of such methodologies:

> a 'strange' space delineated by theatre performances, games, and exercises can surface between otherwise hidden material-affects, making room for critique and, in turn, intervention... [which] justifies an expansion of what counts as research material to include drawn and written notes, audio and visual recordings of theatre practices and theatre-making processes, and other co-produced outputs from development.

By working with participants and the complex set of negotiations that requires, it becomes possible not just to collect and disseminate the stories of those people involved – though the plurality of stories therein present significant challenges in constructing the plot (Raynor 2017a) – but also to materialize a whole host of relations and affects which "become palpable through theatre practice" such that participants' lived realities become "enlivened in performance" (Raynor 2019: 693, 694). For example, Raynor (2017a: 194) shows this through her research which co-wrote/co-produced a fictional play that was informed by and attempted to "'capture and evoke' women's everyday encounters with austerity".

A further example where geographers have experimented in multi-media and/or non-text-based forms of output emerges from scholarship on sound and soundscapes (see Chapter 6). Thus far, "phonography is methodologically underdeveloped in comparison with photographic and videographic techniques" (Gallagher 2015: 564; also see Gallagher and Prior 2014). However, there are some very interesting developments emerging here, in part informed by the ideas and issues that NRTs have brought to geographers' attention. Participant-generated sound diaries have been used in understanding individuals embodied knowledges and relationships with places (Duffy and Waitt 2011). When it comes to presenting such sonic research, Gallagher (2015: 560) illustrates a range of ways in which 'field recordings' – "the production, circulation, and playback of audio recordings of the myriad soundings of the world" – can do "geographical work" beyond that which takes place through academic texts, numbers, imagery, and maps. Gallagher shows how field recordings can be worked with in producing different sorts of research outputs which might engage beyond traditional academic audiences. Such sound recordings might often be seen to have a representational power in (re)producing acoustic knowledge about certain places or environments. However, through their editing and the montage or juxtaposition of various recorded sounds, such recordings can be performative in the way that they combine "something happening here and now as well as a document of another time and place" (Gallagher 2015: 561). Listening to these sound recordings unfolds through time and so must be enacted in its own space-time. The sound works produced from such recordings then have the potential to be affective given the ways in which they draw out the affective power of such representational matters. As Gallagher (2015: 571) reflects:

> "it is precisely this mix of different registers that makes these works 'work'. Their representational aspects are all the more effective for their affects. These affects enable the works to tell compelling stories, to represent places in a way that carries a certain force of truth – partial, messy, and multiplicious, but truth nonetheless; knowledge that hits home. Conversely, the representational aspects of these works, the fact that they are recognisably portraits of people and places, heightens their affects".

Such creative outputs – in the form of plays, soundworks, or other forms of performance art – move some way beyond the realm of traditional academic output and in doing so introduce more of the subject matter of work following in the wake of NRTs. Here, embodied experience, affective relations, performative encounters, and so on become encountered as part of that output itself.

Conclusion

This chapter has provided an introduction to the ongoing efforts being made in light of NRTs to develop methodologies that attend to the challenges such ideas pose to how social science research has commonly been done. Given that much of the initial and ongoing development of NRTs has been more concerned with conceptual matters ahead of how research might be done, this remains a less developed and still emerging set of concerns. However, what is becoming clear is that NRTs do not call for the use of specific methods. It is not simply the case that research in light of NRTs should be (auto-)ethnographic or that video should be used in the research process. Equally, it is not the case that certain methods should not be used.

Interviews, focus groups, and the like potentially still have a place here. What is clear is that it is more about the style of work being pursued and how such methods are understood and used – to what aims, agendas, or ends they are applied and not applied – and so it becomes more a question of what our methodologies look like in light of NRTs. There is little in the way of 'answers', then, in this work; it is more a case of challenging us to think about what questions we pose, how we pose them, and how we respond to the challenges they present us with.

Similarly, such research raises issues over presentation. Parroting critiques of NRTs, how can we go about 'representing the non-representational' in our research outputs? Again, the response comes in reframing the ends to which we aspire and the way we understand the challenges NRTs raise. The concern here is neither about trying to represent things better nor about reveling in such failure in some sort of relativist, nihilistic manner. Rather, it becomes a case of asking how we might become more sensitive to questions over how and what we write and of what that writing tries to do. It might mean we engage in more performative forms of writing which seek to make manifest something of the events and relations we are interested in; less trying to produce a faithful representation of how things are and instead draw out the relations, affects, and becomings that the researcher or researched bore witness to. Or it might mean experimenting with forms of presentation that augment or entirely move away from written forms; trying to engage other senses, move bodies with moving bodies, to make words resound, and so on. Where this all gets to is an open-ended question.

In this chapter, I have tried to stick quite closely to questions of how methodologies and styles of doing research have developed in light of NRTs. However, it's important in concluding to note that, as is increasingly the case in geography today, the boundaries here are becoming increasingly blurred as to where NRTs' influences begin and end. There are a host of parallel, overlapping, complimentary, and subtly different trajectories unfolding in recent geographic work all broadly coming in the wake of the 'practice turn' (see Chapter 1). For example, there is another story to be told here of a 'creative turn' in cultural geography whereby the sorts of attention to artistic styles of working found in the work discussed in this chapter are both mirrored and differ (for an overview, see Hawkins 2014, 2017; and in relation to NRTs specifically, Boyd 2017b). There has been an increased interest in participatory forms of work where creativity is studied by being creative with creatives. As with the auto-ethnographic nature of a lot of work informed by NRTs, here, geographers have collaborated with artists in their practice but also have worked with artists in developing responses to mutually held matters of concern (Hawkins 2015; though see Williams 2017). Further, there is also a story to be told around the 'mobilities turn' in the social sciences whereby novel attempt to research bodies on the move have emerged (see Adey 2017; Merriman 2014). And there are common reference points therein in terms of developments like video methods and go-along approaches (Cook *et al.* 2016; Spinney 2009), auto-ethnographic styles of working (Jones 2005), and creative attempts to present those encounters that unfold in such movement (Latham and Wood 2015). Ultimately, such developments are not silos, completely separated from each other and so the potential for parallel and divergent developments holds the possibility for interesting future opportunities in developing innovative methodologies.

Further reading

Vannini (2015a) is the first book-length engagement with questions of methodology when it comes to NRTs. This book does not offer the sort of 'how to' guide that is common when it comes to edited collections focused around research methods. Rather, it includes a set of broader reflections on the sort of style of work – again, methodology rather than method – that might follow from NRTs and how that work might be presented. The introductory chapter is particularly effective in introducing such concerns (as well as NRTs themselves). **Boyd and Edwardes (2019)** introduce the intersection of geography informed by NRTs and work from and by creative arts/artists. In this, we see reflections on various forms of performance, art practice, dance, theatre, and sound work; the spaces in which performances/productions take shape and unfold; theoretical reflections on NRTs and such practice; and examples of creative writing. This provides the most sustained engagement so far in terms of where NRTs might lead for research production. *cultural geographies* regularly publishes short pieces as part of their 'in Practice' section which experiment with various aspects of doing research in cultural geography. This mixes reflection on different methodologies, styles, or experiences working in the field (including as part of collaborations with artists or other practitioners) and on how these might be presented. This also includes attempts to present geographic research creatively through visual or textual means. This provides a range of useful introductions to how geographic research might look in the wake of NRTs and related developments in geography more broadly.

CONCLUDING

Looking back

When I set out to write this book, there were several things that I did not want to do. Most of all, I didn't want to write (yet another) call for 'NRT' which articulates (a version of) what 'NRT' really is or really should be. There's an act of judgment in that that I simply did not want to engage in and, as the reference list for this book attests, enough of such manifesto-ing has already taken place. As part of that, I didn't want to try to police what 'NRT' is, I didn't want to legislate on what counts as 'NRT' and what doesn't, and I didn't want to have the last word in a series of debates that have taken place at various points in NRTs' (ongoing) development. A book like this will always risk doing some of that or be read as doing that whether it is the intention or not. Hopefully, I've somehow navigated that tension here.

In setting out to write this book, my initial intention was instead to produce a relatively even-keeled introduction to NRTs by discussing work often associated with NRTs and some of the responses to that work. To me, that has been relatively absent from the existing literature developing or commenting on NRTs/'NRT'. Any critique present in that has been presented with the intention of trying to register the productive tensions that NRTs have brought about and continue to produce. For some, that will not be enough of a critical statement or intervention but it is what I am willing to offer here. What that does mean, though, is that this book is (knowingly) one 'take' on NRTs. I'm sure some would question my inclusions and exclusions here. Some might be frustrated that I haven't discussed a certain debate or piece of work, or even point within a piece of work, or even more discussed it in a particular way. Others might be surprised that I have included them or certain work in the amorphous cast of characters who may or may not immediately be included in discussions of NRTs. Others still might disagree with my treatment of certain debates. That is all fair. This is, again, one take on NRTs that emerges from around 15 years of engaging with and contributing to their ongoing development. I'm not pretending to be 'the' authority. Given NRTs' increasing diversity, various off-shots, increasingly blurred boundaries, the expanding cast of advocates, practitioners, critics, and converts, amongst other matters of concern, there are many other ways that a book like this could have taken shape.

Such caveats and qualifications aside, and by way of a brief conclusion, rather than undertake the classic agenda setting gesture of surmising the preceding discussion into a definitive statement on 'what NRT really is' or 'what NRT really should be', I instead want to outline what I think are, based on looking back over the preceding chapters, some key ongoing challenges for the future of NRTs and their continued emergence and differing.

Looking forward

'Performative critique'

One challenge for 'NRT' that has become clear over the course of writing this book is the ways in which critiques have become established through what might be seen as performative citational practices, a performative form of critique. I'm referring here specifically to Judith Butler's account of performativity which emphasizes the reinforcement of norms through the (re)citation of past ideas or expectations (see Chapter 2). Certain (literal) citational practices here do something when it comes to discussion of NRTs. First, the citation of a small number of articles written by one (or a select few) individual(s) commonly occurs here. Second, a small number of interventions which are critical of specific points in 'NRT,' taken from the aforementioned small set of work, come to be regularly cited in subsequent scholarship. This leads to two issues emerging for NRTs. The former point means that select work of that small number of individuals (or a single individual) is taken to be 'NRT' and/or those individuals made synonymous with 'NRT'. The latter point means that the re-citation of such critiques acts to reproduce, reinforce, and legitimise those critiques of 'NRT' which (a) were themselves debated based on their first airing and (b) fix 'NRT', very much in the singular, in time and place (around the decade from 1996 to 2006). Through such citational practice – citing a limited range of work and reciting the critiques made of that work – these critical points can come to be simply accepted without further thought or reflection. They become true through their repetition. Those citing them do not necessarily return to the material being critiqued for themselves and rather take the critiquing authors' word for it. This is really important when it comes to the future of NRTs. In fixing 'NRT' in this way – in the singular, to a specific space-time, and to a few works of a few individuals – such citational practices fail to recognize the ongoing evolution of NRTs.

This is perhaps most evident in the repeated concern with 'the non-representational' and the assumption that NRTs do not take an interest in representations. Regardless of how many times those engaged in the development of NRTs have said and/or shown that NRTs are concerned with representations, early critiques which claimed that NRTs did not attend to representation or denied their importance recur and are taken as correct. The problem that is the name 'non-representational theory' just won't go away. Even more evident though is the recurring critique that work on affect, and related themes, doesn't attend to issues of power or difference. Again, this critique itself is by no means 'true' but the repeated citation of certain critiques of 'NRT' from the early and mid-2000s acts to perpetuate this position. In doing so, such citational practices falsely foreclose NRTs' discussions of affect. Certain articles from the early 2000s come to be taken as definitive statements of what affect is or what 'NRT' means by affect. Work by a host of subsequent scholars (writing after those critiques, whatever their validity) is not read or recognized for the ways it makes clear that such work on affect did and does in increasingly nuanced ways recognize the imbrications of affect, difference, and power. Effectively, we take such critiques as given and in so doing freeze the movement of thought. This acts to shut things down rather than allowing our thinking about how the world is lived to develop.

In a sense, this is a challenge that NRTs will find difficult to respond to so long as such citational practices continue. It is not easy to break outside of such norm-ing. However, as a range of recent work has noted (albeit in different contexts – see Mott and Cockayne (2017)), a concern for citational practices and the politics of citation should allow for a recognition that NRTs are being developed by an increasingly diverse cast of actors who are developing these ideas in nuanced and exciting ways. While a range of 'usual suspects' have featured prominently in the preceding chapters, I hope that this book might also make a small contribution to this issue in showing a range of work by a wider cast of characters that has recently, and continues to, develop NRTs in a range of critical and creative ways.

Philosophy and geography

I made a conscious decision at the outset of writing this book to try to focus on the work being done by geographers and their engagement with a host of ideas drawn from beyond the bounds of that discipline rather than engaging in the exposition of the work of various philosophers and social theorists. I could have easily written a book where each chapter picked an influential philosopher, theorist, school of thought, and so on and introduced their impact on NRTs. Equally, I could have spent a whole lot more time in various boxes and asides introducing such work within this book's structure. That would have led to a very different book.

That question of philosophical/theoretical exposition does raise an ongoing challenge to the broader reception and development of NRTs within geography: namely, that most geographers aren't trained in philosophy. Geographers have evidently read the work of philosophers for some time; it's nothing really all that new. However, there is a risk here that such engagements devolve into highly nuanced and technical debates over, for example, what the proper way to understand affect is, or what the best way of conceptualizing the event is, or how it really is that subjects emerge in practice. Wrestling with such questions – and the very significant academic labor that entails – can be a worthwhile exercise. But I'd suggest that there needs to be a clear awareness amid that of the 'so what?'; what does an engagement with such ideas do in terms of helping geographers/geography better understand the functioning of various situations, relations, circumstances, or events in the world. And there is a lot of scope to think about those implications today in a world filled with fake news, hateful populist politics, algorithmic intelligence, increasingly networked media technologies, increasingly palpable collectively felt materialities and atmospheres, extensive augmentations of bodies through banal and spectacular prosthesis, and so on. Without thinking in 'wordly' terms, such theoretical debating risk taking place within little more an echo chamber occupied by a select few.

This isn't a naïve anti-intellectual call against theory or theoretical nuance/exactness. Rather, to me, we need to be attentive to the conjunction philosophy *and* geography. Focusing too much one side, or an either/or within that, will limit – to put it crassly – the use of NRTs and the potential spread of these ideas more broadly amongst those that scholarship succeeds in engaging. These ideas would have less impact on our understanding and action in the world. There is, to me at least, something significant going on in this work, and we need to take care in ensuring that significance is made clear to a wide audience.

One or many politics

Probably, the most common recurring point of critique which has emerged throughout reading for and writing this book has related to what sort of politics is or isn't present in work influenced by NRTs. The very recurrence of that makes me think that it is something that will not go away anytime soon (though it might continue to mellow in tone). It has become clear that many of NRTs' conceptual, thematic, and stylistic leanings have caused ripples in geography around a host of dearly held concerns. Matters of identity politics, radical action, critique, situated knowledge, and the like have been flagged as things that 'NRT' does not engage with. Again, returning to the problematic of naming, the refrain of NRTs' ongoing development may well have been the drawing of a distinction between 'the representational' and the 'non-representational' or that 'NRT' doesn't do representation.

Again, I hope it's clear from the preceding pages that this critique has been heard and responded to in a host of ways in the ongoing development of NRTs in geography. That comes both in the retorts of 'but it's not about forgetting representations' or 'NRTs look at what representations do', but also in work which *shows* how affects, practices, relations, events, and the like unfold in ways which are in some sense contextual or circumstantial. Such contexts or circumstances are not necessarily determinative of how an event, practice, encounter or whatever will unfold and we don't necessarily know in advance what impact they may or may not have. The key point is that they hold the potential to impact upon the unfolding of social life and there is a politics associated with this. The situated nature of bodies and materialities is increasingly evident in work engaged with NRTs, and it's simply being done in ways that depart from how past geographers might have thought about them.

And that gets to the key point here: how accepted such shifts in thinking are and how much the discipline is willing to accommodate those differences. Despite the exuberant tenor of Thrift's early outline and some of the strident debate that has unfolded since, I've never seen NRTs as being presented to geography as the way geographic research must be done, as doctrinal, or as something everyone must convert to. I can see how some might have felt that, but I'm not sure it ever was really the intention. My sense is more that it was a process of negotiation, starting with a more radical opening offer knowing that there was ground to be given. Taken in that sense NRTs have presented, in equal parts, something to trouble norms and something that might inspire thought. If that troubling leads to a clarification of what researchers hold important and an additionally nuanced and consciously reflected upon engagement with that, then that can only be a good thing. If that means that the world all of a sudden appears to us differently and that new terrains and new questions open up, then that can only be a good thing also. In either case, it's about challenging geography's taken-for-granted ways of thinking about doing.

Lines of flight?

This book is probably a decade late. Someone really should have written it before me and before now. I think it probably would have been a much easier task to complete when NRTs were easier to bound and less diffuse in their influence and spread. That book probably could have been written in the singular without troubling over the grammatical challenges of talking of 'NRTs'. It would have been more straightforward to decide who to include or exclude in that discussion. It would have been several tens of thousands of words shorter.

This all does, though, highlight a point of tension (and potential success) in terms of the development of NRTs in geography at the point in time that this book was written. The ongoing profusion of themes, ideas, concepts, styles, presentational challenges, and the like that has taken place in geography in light of NRTs has meant that 'NRT' as a singular thing is increasingly hard to see. NRT or NRTs is/are not necessarily something that is used to frame a host of work which clearly follows in its/their wake. Certain concepts largely introduced into geography by Thrift and others in the initial explorations of NRTs have supplanted that framing. Affect is probably the most obvious case study for this, which is itself already morphing into discussions of a host of 'atmospheres'. That said, a more diffuse emphasis on practice and agential materialities also appears to be an ongoing legacy here.

This book will, in all likelihood, act to draw bounds and retain a consistency amongst this work. The title is in the singular, after all. This book might equally, though, end up being something more like a eulogy for some thing – a movement, a moment, a feeling, a confluence of energies – that has now passed. That's not to say that the ideas, agendas, and arguments covered in the preceding 90,000 or so words are now passé or irrelevant. Rather, it's to say that what was initially introduced into geography as 'non-representational thought', and which later reluctantly became 'NRT', is now, and will continue to become, something different, something less coherent, something more than it currently is or might be. There may remain some identity amid this difference and differing, allowing for 'NRT' to remain a meaningful, if limiting, shorthand. But the extent to which that shorthand will ever accurately account for what is going on here, and continues to go on, is far less likely.

REFERENCES

Abrahamsson, C. and Abrahamsson, S. (2007) In conversation with a body conveniently known as Stelarc. *cultural geographies*, 14, pp. 293–308.

Abrahamsson, S., Bertoni, F., Mol, A. and Martin, R.I. (2015) Living with omega-3: New materialism and enduring concerns. *Environment and Planning D: Society and Space*, 33, pp. 4–19.

Adey, P. (2009) Facing airport security: Affect, biopolitics, and the preemptive securitisation of the mobile body. *Environment and Planning D: Society and Space*, 27, pp. 274–295.

Adey, P. (2013) Air/atmospheres of the megacity. *Theory, Culture & Society*, 30, pp. 291–308.

Adey, P. (2014) Security atmospheres or the crystallisation of worlds. *Environment and Planning D: Society and Space*, 32, pp. 834–851.

Adey, P. (2015) Air's affinities: Geopolitics, chemical affect and the force of the elemental. *Dialogues in Human Geography*, 5, pp. 54–75.

Adey, P. (2017) *Mobility* (2nd ed). Routledge, London.

Adey, P., Anderson, B. and Guerrero, L.L. (2011) An ash cloud, airspace and environmental threat. *Transactions of the Institute of British Geographers*, 36, pp. 338–343.

Adey, P., Brayer, L., Masson, D., Murphy, P., Simpson, P. and Tixier, N. (2013) 'Pour votre tranquillité': Ambiance, atmosphere, and surveillance. *Geoforum*, 49, pp. 209–309.

Ahmed, S. (2010) Happy objects. In Gregg, M. and Seigworth, G.J. (eds) *The Affect Theory Reader*. Duke University Press, London.

Amin, A. (2006) The good city. *Urban Studies*, 43, pp. 1009–1023.

Amin, A. and Thrift, N. (2002) *Cities: Reimaging the Urban*. Polity Press, Cambridge.

Amin, A. and Thrift, N. (2013) *Arts of the Political: New Openings for the Left*. Duke University Press, London.

Anderson, B. (2004) Time-stilled space-slowed: How boredom matters. *Geoforum*, 35, pp. 739–754.

Anderson, B. (2005) Practices of judgement and domestic geographies of affect. *Social & Cultural Geography*, 6, pp. 645–660.

Anderson, B. (2006) Becoming and being hopeful: Towards a theory of affect. *Environment and Planning A*, 24, pp. 733–752.

Anderson, B. (2007) Hope for nanotechnology: Anticipatory knowledge and the governance of affect. *Area*, 39, pp. 156–165.

Anderson, B. (2009a) Non-representational theory. In Gregory, D., Johnston, R., Pratt, G., Watts, M.J. and Whatmore, S. (eds) *The Dictionary of Human Geography*. Wiley-Blackwell, Oxford.

Anderson, B. (2009b) Affective atmospheres. *Emotion, Space and Society*, 2, pp. 77–81.

Anderson, B. (2014) *Encountering Affect: Capacities, Apparatuses, Conditions*. Ashgate, Aldershot.

Anderson, B. (2017) "We will win again. We will win a lot": The affective styles of Donald Trump. Available at: http://societyandspace.org/2017/02/28/we-will-win-again-we-will-win-a-lot-the-affective-styles-of-donald-trump/

Anderson, B. (2019) Cultural geography II: The force of representations. *Progress in Human Geography*, 43, pp. 1120–1132.

Anderson, B. and Ash, J. (2015) Atmospheric methods. In Vannini, P. (ed) *Non-representational Methodologies: Re-envisioning Research*. Routledge, London.

Anderson, B. and Harrison, P. (2006) Questioning affect and emotion. *Area*, 38, pp. 333–335.

Anderson, B. and Harrison, P. (2010a) The promise of non-representational theories. In Anderson, B. and Harrison, P. (eds) *Taking-place: Non-representational Theories and Geography*. Ashgate, Farnham.

Anderson, B. and Harrison, P. (eds) (2010b) *Taking-place: Non-representational Theories and Geography*. Ashgate, Aldershot.

Anderson, B., Kearnes, M. and Doubleday, R. (2007) Geographies of nano-technoscience. *Area*, 39, pp. 139–142.

Anderson, B., Keanes, M., McFarlane, C. and Swanton, D. (2012) On assemblages and geography. *Dialogues in Human Geography*, 2, pp. 171–189.

Anderson, B. and McFarlane, C. (2011) Assemblage and geography. *Area*, 43, pp. 124–127.

Anderson, B., Morton, F. and Revill, G. (2005) Editorial: Practices of music and sound. *Social & Cultural Geography*, 6, pp. 638–644.

Anderson, B. and Tolia-Kelly, D.P. (2004) Matter(s) in social and cultural geography. *Geoforum*, 35, pp. 669–674.

Anderson, B. and Wylie, J. (2009) On geography and materiality. *Environment and Planning A*, 41, pp. 318–335.

Anderson, J. (2010) *Understanding Cultural Geography: Places and Traces*. Routledge, London.

Andrews, G. (2018) *Non-representational Theory & Health: The Health in Life in Space-time Revealing*. Routledge, Abingdon.

Ash, J. (2010a) Architectures of affect: Anticipating and manipulating the event in processes of videogame design and testing. *Environment and Planning D: Society and Space*, 28, pp. 653–671.

Ash, J. (2010b) Teleplastic technologies: Charting practices of orientation and navigation in videogaming. *Transactions of the Institute of British Geographers*, 35, pp. 414–430.

Ash, J. (2013) Rethinking affective atmospheres: Technology, perturbation and space times of the non-human. *Emotion, Space and Society*, 49, pp. 20–28.

Ash, J. (2015a) *The Interface Envelope: Gaming, Technology, Power*. Bloomsbury Press, New York.

Ash, J. (2015b) Technology and affect: Towards a theory of inorganically organised objects. *Emotion, Space and Society*, 14, pp. 84–90.

Ash, J. (2016) Theorizing studio space: Spheres and atmospheres in a video game design studio. In Farias, I. and Wilkie, A. (eds) *Studio Studies: Operations, Topologies & Displacements*. Routledge, London.

Ash, J. and Gallacher, L.A. (2011) Cultural geography and videogames. *Geography Compass*, 5/6, pp. 351–368.

Ash, J. and Simpson, P. (2016) Geography and post-phenomenology. *Progress in Human Geography*, 40, pp. 48–66.

Ash, J. and Simpson, P. (2019) Postphenomenology and method: Styles for thinking the (non)human. *Geohumanities*, 5, pp. 139–156.

Auslander, P. (1999) *Liveness: Performance in a Mediatized Culture*. Routledge, London.

Barnes, J. and Duncan, J. (eds) (1992) *Writing Worlds: Discourse, Text and Metaphor in the Representation of Landscape*. Routledge, London.

Barnett, C. (2004) A critique of the cultural turn. In Duncan, J.S., Johnson, N.C. and Schein, R.H. (eds) *A Companion to Cultural Geography*. Blackwell, Oxford.

Barnett, C. (2008) Political affects in public space: Normative blind-spots in non-representational ontologies. *Transactions of the Institute of British Geographers*, 33, pp. 186–200.

Barnett, C. (2009) Cultural turn. In Gregory, D., Johnston, R., Pratt, G., Watts, M.J. and Whatmore, S. (eds) *The Dictionary of Human Geography*. Wiley-Blackwell, Oxford.

Barron, A. (2019) More-than-representational approaches to the life-course. *Social & Cultural Geography*, Early Online at https://doi.org/10.1080/14649365.2019.1610486

Bennett, J. (2004) The force of things: Steps towards an ecology of matter. *Political Theory*, 32, pp. 347–372.

Bennett, J. (2010) *Vibrant Matter: A Political Ecology of Things*. Duke University Press, Durham.

Berlant, L. (2007) Unfeeling Kerry. *Theory and Event*, 8.

Bille, M., Bjerregaard, P. and Sorensen, T.F. (2015) Staging atmospheres: Materiality, culture, and the texture of the in-between. *Emotion, Space and Society*, 15, pp. 31–38.

Billig, M. (1995) *Banal Nationalism*. Sage, London.

Bingham, N. (1996) Object-ions: From technological determinism towards geographies of relations. *Environment and Planning D: Society and Space*, 14, pp. 635–657.

Bingham, N. (2006) Bees, butterflies, and bacteria: Biotechnology and the politics of nonhuman friendship. *Environment and Planning A*, 38, pp. 483–498.

Bingham, N. and Thrift, N. (2000) Some new instructions for travelers: The geography of Bruno Latour and Michel Serres. In Crang, M. and Thrift, N. (eds) *Thinking Space*. Routledge, London.

Bissell, D. (2007) Animating suspension: Waiting for mobilities. *Mobilities*, 2, pp. 277–298.

Bissell, D. (2008) Comfortable bodies: Sedentary affects. *Environment and Planning A*, 40, pp. 1697–1712.

Bissell, D. (2009a) Conceptualising differently-mobile passengers: Geographies of everyday encumbrance in the railway station. *Social & Cultural Geography*, 10, pp. 173–195.

Bissell, D. (2009b) Travelling vulnerabilities: Mobile timespaces of quiescence. *cultural geographies*, 16, pp. 427–445.

Bissell, D. (2009c) Inconsequential materialities: The movements of lost effects. *Space and Culture*, 12, pp. 95–115.

Bissell, D. (2010) Passenger mobilities: Affective atmospheres and the sociality of public transport. *Environment and Planning D: Society and Space*, 28, pp. 270–289.

Bissell, D. (2011) Thinking habits for uncertain subjects: Movement, stillness, susceptibility. *Environment and Planning A*, 43, pp. 2649–2665.

Bissell, D. (2013) Habit misplaced: The disruption of skilful performance. *Geographical Research*, 51, pp. 120–129.

Bissell, D. (2015) Virtual infrastructures of habit: The changing intensities of habit through gracefulness, restlessness and clumsiness. *cultural geographies*, 22, pp. 127–146.

Blacksell, M. (2005) Comment – A walk on the South West Coast Path: A view from the other side. *Transactions of the Institute of British Geographers*, 30, pp. 518–520.

Bogost, I. (2012) *Alien Phenomenology, or What It's Like to Be a Thing*. Minnesota University Press, Minneapolis.

Bohme, G. (2017) *The Aesthetics of Atmosphere* (edited by Thibaud, J.-P.). Routledge, London.

Bondi, L. (2005) Making connections and thinking through emotions: Between geography and psychotherapy. *Transactions of the Institute of British Geographers*, 30, pp. 433–448.

Boyd, C.P. (2017a) Research poetry and the non-representational. *ACME: An International Journal for Critical Geographies*, 16, pp. 210–223.

Boyd, C.P. (2017b) *Non-representational Geographies of Therapeutic Art Making: Thinking through Practice*. Palgrave Macmillan, London.

Boyd, C.P. and Edwardes, C. (eds) (2019) *Non-representational Theory and the Creative Arts*. Palgrave MacMillan, London.

Brennan, T. (2004) *The Transmission of Affect*. Cornell University Press, Ithaca.

Brigstocke, J. (2012) Defiant laughter: Humour and the aesthetics of place in late 19th century Montmartre. *cultural geographies*, 19, pp. 217–235.

Brigstocke, J. (2014) *The Life of the City: Space, Humour, and the Experience of Truth in Fin-de-siècle Montmartre*. Ashgate, Aldershot.

Brigstocke, J. and Noorani, T. (2016) Posthuman attunements: Aesthetics, authority and the arts of creative listening. *GeoHumanities*, 2, pp. 1–7.

Brown, S.S. and Mathewson, K. (1999) Sauer's descent?: Or Berkeley roots forever? *Yearbook of the Association of Pacific Coast Geographers*, 61, pp. 137–157.

Bull, M. (2000) *Sounding Out the City: Personal Stereos and the Management of Everyday Life*. Berg, Oxford.

Butler, J. (1990) Performative acts and gender constitution: An essay in phenomenology and feminist theory. In Case, S.-E. (ed) *Performing Feminism: Feminist Critical Theory and Theatre*. Johns Hopkins University Press, Baltimore.

Butler, J. (1993) *Bodies That Matter: On the Discursive Limits of 'Sex'*. Routledge, London.

Butler, J. (1999) *Gender Trouble: Feminism and the Subversion of Identity*. Routledge, London.

Buttimer, A. (1976) Grasping the dynamism of the lifeworld. *Annals of the Association of American Geographers*, 66, pp. 277–292.

Buttimer, A. and Seamon, D. (eds) (1980) *The Human Experience of Space and Place*. Croom Helm, London.

Cadman, L. (2009) Nonrepresentational theory/nonrepresentational geographies. In Kitchin, R. and Thrift, N. (eds) *International Encyclopedia of Human Geography*. Elsevier, Amsterdam.

Carney, G.O. (1998) Music geography. *Journal of Cultural Geography*, 18, pp. 1–10.

Carney, G.O. (ed) (2003) *The Sounds of People and Places: A Geography of American Music from Country to Classical and Blues to Bop* (4th ed). Rowman & Littlefield, Lanham.

Carter, S.R. and McCormack, D.P. (2006) Film, geopolitics and the affective logics of intervention. *Political Geography*, 25, pp. 228–245.

Castree, N. and MacMillan, T. (2004) Old news: Representation and academic novelty. *Environment and Planning A*, 35, pp. 469–480.

Cloke, P. and Jones, O. (2001) Dwelling, place, and landscape: An orchard in Somerset. *Environment and Planning A*, 33, pp. 649–666.

Closs-Stephens, A. (2016) The affective atmospheres of nationalism. *cultural geographies*, 23, pp. 181–198.

Colls, R. (2007) Materialising bodily matter: Intra-action and the embodiment of 'Fat'. *Geoforum*, 38, pp. 353–365.

Colls, R. (2012) Feminism, bodily difference and non-representational geographies. *Transactions of the Institute of British Geographers*, 37, pp. 430–445.

Connell, C. and Gibson, C. (2003) *Sound Tracks: Popular Music, Identity and Place*. London, Routledge.

Connolly, W. (2002) *Neuropolitics*. University of Minnesota Press, Minneapolis.

Cook, I.J. and Crang, P. (1996) 'The world on a plate': Culinary culture, displacement and geographical knowledges. *Journal of Material Culture*, 1, pp. 131–153.

Cook, S., Shaw, J. and Simpson, P. (2016) Jography: Exploring meanings, experiences and spatialities of road-running. *Mobilities*, 11, pp. 744–769.

Coole, D. and Frost, S. (2010) Introducing the new materialisms. In Coole, D. and Frost, S. (eds) *New Materialisms: Ontology, Agency, and Politics*. Duke University Press, Durham.

Cosgrove, D. (1984) *Social Formation and Symbolic Landscape*. University of Wisconsin Press, Madison.

Cosgrove, D. and Daniels, S. (eds) (1988) *The Iconography of Landscape*. Cambridge University Press, Cambridge.

Cosgrove, D. and Jackson, P. (1987) New directions in cultural geography. *Area*, 19, pp. 95–101.

Crang, M. (1994) It's showtime: On the workplace geographies of display in a restaurant in southeast England. *Environment and Planning D: Society and Space*, 12, pp. 675–704.

Crang, M. (2001) Rhythms of the city: Temporalised space and motion. In May, J. and Thrift, N. (eds) *Timespace: Geographies of Temporality*. Routledge, London.

Crang, M. (2003) Qualitative methods: Touchy, feely, look-see? *Progress in Human Geography*, 27, pp. 494–504.

Cresswell, T. (2003) Landscape and the obliteration of practice. In Anderson, K., Domosh, M., Pile, S. and Thrift, N. (eds) *The Handbook of Cultural Geography*. Sage, London.

Cresswell, T. (2006a) 'You cannot shake that shimmie here': Producing mobility on the dance floor. *cultural geographies*, 13, pp. 55–77.

Cresswell, T. (2006b) *On the Move: Mobility in the Modern Western World*. Routledge, London.

Cresswell, T. (2012) Nonrepresentational theory and me: Notes of an interested sceptic. *Environment and Planning D: Society and Space*, 30, pp. 96–105.

Cresswell, T. (2013a) *Geographic Thought: A Critical Introduction*. Wiley-Blackwell, Oxford.

Cresswell, T. (2013b) Soil. Penned in the Margins, London.

Daniels, S. (1989) Marxism, culture, and the duplicity of landscape. In Peet, R. and Thrift, N. (eds) New Models in Geography: Volume Two. Unwin Hyman, London.

Daniels, S. and Cosgrove, D. (eds) (1989) The Iconography of Landscape. Cambridge University Press, Cambridge.

Daniels, S. and Lorimer, H. (2012) Until the end of days: Narrating landscape and environment. cultural geographies, 19, pp. 3–9.

Davies, G. and Dwyer, C. (2007) Qualitative methods: Are you enchanted or are you alienated? Progress in Human Geography, 31, pp. 257–266.

DeLanda, M. (2006) A New Philosophy of Society: Assemblage Theory and Social Complexity. Continuum, London.

Deleuze, G. (1978) Seminar given on Spinoza on 24/01/1978. http://www.web-deleuze.com/php/texte/php?cle=14&groupe=Spinoza&langue=2

Deleuze, G. (1988) Spinoza: Practical Philosophy. City Lights Books, San Francisco.

Deleuze, G. (2004) The Logic of Sense. Continuum, London.

Deleuze, G. and Guattari, F. (2004) A Thousand Plateaus. Continuum, London.

Deleuze, G. and Parnet, C. (2006) Dialogues II. Continuum, London.

DeSilvey, C. (2006) Observed decay: Telling stories with mutable things. Journal of Material Culture, 11, pp. 318–338.

Dewsbury, J.-D. (2000) Performativity and the event: Enacting a philosophy of difference. Environment and Planning D: Society and Space, 18, pp. 473–496.

Dewsbury, J.-D. (2003) Witnessing space: 'Knowledge without contemplation'. Environment and Planning A, 35, pp. 1907–1932.

Dewsbury, J.-D. (2007) Unthinking subjects: Alain Badiou and the event of thought in thinking politics. Transactions of the Institute of British Geographers, 32, pp. 443–459.

Dewsbury, J.-D. (2009) Affect. In Kitchin, R. and Thrift, N. (eds) International Encyclopedia of Human Geography. Elsevier, Amsterdam.

Dewsbury, J.-D. (2010a) Performative, non-representational, and affect-based research: Seven injunctions. In Delyser, D., Aitken, S., Craig, M., Herbert, S. and McDowell, L. (eds) The SAGE Handbook of Qualitative Geography. Sage, London.

Dewsbury, J.-D. (2010b) Language and the event: The unthought of appearing worlds. In Anderson, B. and Harrison, P. (eds) Taking-place: Non-representational Theories and Geography. Ashgate, Farnham.

Dewsbury, J.-D. (2011a) Dancing: The secret slowness of the fast. In Cresswell, T. and Merriman, P. (eds) Geographies of Mobilities: Practices, Spaces, Subjects. Ashgate, Aldershot.

Dewsbury, J.-D. (2011b) The Deleuze-Guattarian assemblage: Plastic habits. Area, 43, pp. 148–153.

Dewsbury, J.-D. (2015) Non-representational landscapes and the performative affective forces of habit: From 'Live' to 'Blank'. cultural geographies, 22, pp. 29–47.

Dewsbury, J.-D. and Bissell, D. (2015) Habit geographies: The perilous zones in the life of the individual. cultural geographies, 22, pp. 21–28.

Dewsbury, J.D. and Cloke, P. (2009) Spiritual landscapes: Existence, performance, and immanence. Social & Cultural Geography, 10, pp. 695–711.

Dewsbury, J.-D., Harrison, P., Rose, M. and Wylie, J. (2002) Introduction: Enacting geographies. Geoforum, 33, pp. 437–440.

Dewsbury, J.-D. and Naylor, S. (2002) Practising geographical knowledge: Fields, bodies and dissemination. *Area*, 34, pp. 253–260.

Dixon, D.P. and Jones, J.P. (2004) Poststructuralism. In Duncan, J.S., Johnson, N.C. and Schein, R.H. (eds) *A Companion to Cultural Geography*. Blackwell, Oxford.

Doel, M.A. (2010) Representation and difference. In Anderson, B. and Harrison, P. (eds) *Taking-place: Non-representational Theories and Geography*. Ashgate, Aldershot.

Doel, M.A. and Clarke, D.B. (2007) Afterimages. *Environment and Planning D: Society and Space*, 25, pp. 890–910.

Dolphijn, R. and van der Tuin, R. (2012) *New Materialism: Interviews and Cartographies*. Open Humanities Press, Ann Arbor.

Doughty, K., Duffy, M. and Harada, T. (2016) Practices of emotional and affective geographies of sound. *Emotion, Space and Society*, 20, pp. 39–41.

Dowling, R., Lloyd, K. and Suchet-Pearson, S. (2018) Qualitative methods 3: Experimenting, picturing, sensing. *Progress in Human Geography*, 42, pp. 779–788.

Duffy, M. and Waitt, G. (2011) Sound diaries: A method for listening to place. *Aether: The Journal of Media Geography*, VII, pp. 119–136.

Duffy, M. and Waitt, G. (2013) Home sounds: Experiential practices and performativities of hearing and listening. *Social & Cultural Geography*, 14, pp. 466–481.

Duffy, M., Waitt, G. and Harada, T. (2016) Making sense of sound: Visceral sonic mapping as a research tool. *Emotion, Space and Society*, 20, pp. 49–57.

Duncan, J. (1990) *The City as Text: The Politics of Landscape Interpretation in the Kandyan Kingdom*. Cambridge University Press, Cambridge.

Duncan, J. (2000) Landscape. In Johnston, R.J., Gregory, D., Pratt, G. and Watts, M. (eds) *The Dictionary of Human Geography* (4th ed). Blackwell, Oxford.

Duncan, J. and Duncan, N. (1988) (Re)reading the landscape. *Environment and Planning D: Society and Space*, 6, pp. 117–126.

Duncan, J. and Duncan, N. (1992) Ideology and bliss: Roland Barthes and the secret histories of landscape. In Barnes, T.J. and Duncan, J. (eds) *Writing Worlds: Discourse, Text and Metaphor in the Representation of Landscape*. Routledge, London.

Dyck, I. and Kearns, R.A. (2006) Structuration theory: Agency, structure and everyday life. In Aitken, S. and Valentine, G. (eds) *Approaches to Human Geography*. Sage, London.

Dyson, F. (2009) *Sounding New Media: Immersion and Embodiment in the Arts and Culture*. University of California Press, London.

Edensor, T. (2000) Walking in the countryside: Reflexivity, embodied practices and ways to escape. *Body and Society*, 6, pp. 81–96.

Edensor, T. (2005) *Industrial Ruins: Space, Aesthetics and Materiality*. Berg, Oxford.

Edensor, T. (2008) Mundane hauntings: Commuting through the phantasmagoric working-class spaces of Manchester, England. *cultural geographies*, 15, pp. 313–333.

Edensor, T. (ed) (2010) *Geographies of Rhythm: Nature, Place, Mobilities and Bodies*. Routledge, London.

Edensor, T. (2012) Illuminated atmospheres: Anticipating and reproducing the flow of affective experience in Blackpool. *Environment and Planning D: Society and Space*, 30, pp. 1103–1122.

Edensor, T. (2015) Producing atmospheres at the match: Fan cultures, commercialisation and mood management in English football. *Emotion, Space, and Society*, 15, pp. 82–89.

Edensor, T. and Sumartojo, S. (2015) Designing atmospheres: Introduction to special issue. *Visual Communication*, 14, pp. 251–265.

Elden, S. (2004a) Rhythmanalysis: An introduction. In Lefebvre, H. *Rhythmanalysis: Space, Time and Every-day Life*. Continuum, London.

Elden, S. (2004b) *Understanding Lefebvre*. Continuum, London.

Emmerson, P. (2017) Thinking laughter beyond humour: Atmospheric refrains and ethical indeterminacies in spaces of care. *Environment and Planning A*, 49, pp. 2082–2098.

Engelmann, S. (2015a) More-than-human affinitive listening. *Dialogues in Human Geography*, 5, pp. 76–79.

Engelmann, S. (2015b) Towards a poetics of air: Sequencing and surfacing breath. *Transactions of the Institute of British Geographers*, 40, pp. 430–444.

Engelmann, S. and McCormack, D.P. (2018) Elemental aesthetics: On artistic experiments with solar energy. *Annals of the American Association of Geographers*, 108, pp. 241–259.

Entrikin, N.J. and Tepple, J.H. (2006) Humanism and democratic place-making. In Aitken, S. and Valentine, G. (eds) *Approaches to Human Geography*. Sage, London.

Fannin, M. (2011) Personal stem cell banking and the problem with property. *Social & Cultural Geography*, 12, pp. 339–356.

Fannin, M., Jackson, M., Crang, P., Katz, C., Larsen, S., Tolia-Kelly, D.P. and Stewart, K. (2010) Author meets critics: A set of review and a response. *Social & Cultural Geography*, 11, pp. 921–931.

Feigenbaum, A. and Kanngieser, A. (2015) For a politics of atmospheric governance. *Dialogues in Human Geography*, 5, pp. 80–84.

Finn, J. (2011) Introduction: On music and movement…. *Aether: The Journal of Media Geography*, VII, pp. 1–11.

Ford, L. (1971) Geographic factors in the origin, evolution, and diffusion of rock and roll music. *Journal of Geography*, 70, pp. 455–464.

Gagen, E.A. (2004) Making America flesh: Physicality and nationhood in early twentieth-century physical education reform. *cultural geographies*, 11, pp. 417–442.

Gallagher, M. (2015) Field recording and the sounding of spaces. *Environment and Planning D: Society and Space*, 33, pp. 560–576.

Gallagher, M. (2016) Sound as affect: Difference, power and spatiality. *Emotion, Space and Society*, 20, pp. 42–48.

Gallagher, M. and Prior, J. (2014) Sonic geographies: Exploring phonographic methods. *Progress in Human Geography*, 38, pp. 267–284.

Gallagher, M., Kanngieser, A. and Prior, J. (2017) Listening geographies: Landscape, affect and geotechnologies. *Progress in Human Geography*, 41, pp. 618–637.

Garrett, B. (2010) Videographic geographies: Using digital video for geographic research. *Progress in Human Geography*, 35, pp. 521–541.

Greenhough, B. (2010) Vitalist geographies: Life and the more-than human. In Anderson, B. and Harrison, P. (eds) *Taking-place: Non-representational Theories and Geography*. Ashgate, Farnham.

Greenhough, B. (2011) Assembling an island laboratory. *Area*, 43, pp. 134–138.

Gregory, D. (2009) Structuration theory. In Gregory, D., Johnston, R., Pratt, G., Watts, M.J. and Whatmore, S. (eds) *The Dictionary of Human Geography*. Wiley-Blackwell, Oxford.

Gregson, N. (2007) *Living with Things: Ridding, Accommodation, Dwelling*. Sean Kingston Publishing, Wantage.

Gumbrecht, H.G. (2004) *Production of Presence: What Meaning Cannot Convey*. Stanford University Press, Stanford.

Haggett, P., Hoare, T. and Jones, K. (2009) *Geography at the University of Bristol*. School of Geographical Sciences, Bristol.

Harrison, P. (2000) Making sense: Embodiment and the sensibilities of the everyday. *Environment and Planning D: Society and Space*, 18, pp. 497–517.

Harrison, P. (2006) Poststructuralist theories. In Aitken, S. and Valentine, G. (eds) *Approaches to Human Geography*. Sage, London.

Harrison, P. (2007) The space between us: Opening remarks on the concept of dwelling. *Environment and Planning D: Society and Space*, 25, pp. 625–647.

Harrison, P. (2008) Corporeal remains: Vulnerability, proximity, and living on after the end of the world. *Environment and Planning A*, 40, pp. 423–445.

Harrison, P. (2009) In the absence of practice. *Environment and Planning D: Society and Space*, 27, pp. 987–1009.

Harrison, P. (forthcoming) A love whereof non-shall speak: Reflections on naming; 'non-representational theory'. In Bissell, D., Rose, M. and Harrison, P. (eds) *Negative Geographies*. University of Nebraska Press, Lincoln.

Harrison-Pepper, S. (1990) *Drawing a Circle in the Square: Street Performing in New York's Washington Square Park*. University Press of Mississippi, London.

Harvey, D. (1982) *The Limits to Capital*. Blackwell Publishers, Oxford.

Harvey, D. (1996) *Justice, Nature and the Geography of Difference*. Blackwell Publishers, Oxford.

Hawkins, H. (2014) *For Creative Geographies: Geography, Visual Arts and the Making of Worlds*. Routledge, London.

Hawkins, H. (2015) Creative geographic methods: Knowing, representing, intervening. On composing place and page. *cultural geographies*, 22, pp. 247–268.

Hawkins, H. (2017) *Creativity*. Routledge, London.

Hensley, S. (2010) Rumba and rhythmic 'natures' in Cuba. In Edensor, T. (ed) *Geographies of Rhythm: Nature, Place, Mobilities and Bodies*. Ashgate, Aldershot.

Hinchliffe, S.J. (2007) *Geographies of Nature*. London, Sage.

Hinchliffe, S.J. (2010) Working with multiples: A non-representational approach to environmental issues. In Anderson, B. and Harrison, P. (eds) *Taking-place: Non-representational Theories and Geography*. Ashgate, Farnham.

Hinchliffe, S.J., Kearnes, M.B., Degen, M. and Whatmore, S. (2005) Urban wild things: A cosmopolitical experiment. *Environment and Planning D: Society and Space*, 23, pp. 643–658.

Hitchen, E. (2016) Living and feeling the austere. *New Formations*, 87, pp. 102–118.

Hitchen, E. (2019) The affective life of austerity: Uncanny atmospheres and paranoid temporalities, *Social & Cultural Geography*, Available at: https://doi.org/10.1080/1464 9365.2019.1574884.

Hitchings, R. (2003) People, plants and performance: On actor network theory and the material pleasures of the private garden. Social & Cultural Geography, 4, pp. 99–114.

Hitchings, R. (2012) People can talk about their practices. Area, 44, pp. 61–67.

Hoelscher, S. and Alderman, D.H. (2004) Memory and place: Geographies of a critical relationship. Social & Cultural Geography, 5, pp. 347–355.

Holloway, J. (2006) Enchanted spaces: The séance, affect and geographies of religion. Annals of the Association of American Geographers, 96, pp. 182–187.

Holloway, J. and Kneale, J. (2008) Locating haunting: A ghost-hunter's guide. cultural geographies, 15, pp. 297–312.

Honeybun-Arnolda, E. (2019) The promise and practice of spontaneous prose. cultural geographies, 26, pp. 395–400.

Horton, J. and Kraftl, P. (2014) Cultural Geographies: An Introduction. Routledge, Abingdon.

Howard, P., Thompson, I. and Waterton, E. (eds) (2018) The Routledge Companion to Landscape Studies (2nd ed). Routledge, London.

Hubbard, P. (2006) City. Routledge, London.

Ingold, T. (2000) The Perception of the Environment: Essays on Livelihood, Dwelling and Skill. Routledge, London.

Ingold, T. (2007) Earth, sky, wind, and weather. Journal of the Royal Anthropological Institute (N.S.), 13, pp. S19–S38.

Ingold, T. (2011) Being Alive: Essays on Movement, Knowledge and Description. Routledge, London.

Ingold, T. (2015) The Life of Lines. Routledge, London.

Jackson, M. and Fannin, M. (2011) Letting geography fall where it may – Aerographies address the elemental. Environment and Planning D: Society and Space, 29, pp. 435–444.

Jackson, P. (1989) Maps of Meaning: An Introduction to Cultural Geography. Unwin Hyman, London.

Jackson, P. (2000) Rematerializing social and cultural geography. Social & Cultural Geography, 1, pp. 9–14.

Jacobs, J.M. and Nash, C. (2003) Too little, too much: Cultural feminist geographies. Gender, Place and Culture: A Journal of Feminist Geography, 10, pp. 265–279.

Jayne, M., Valentine, G. and Holloway, S.L. (2008) Geographies of alcohol, drinking and drunkenness: A review of progress. Progress in Human Geography, 32, pp. 247–263.

Johnson, N.C. (2004) Public memory. In Duncan, J.S., Johnson, N.C. and Schien, R.H. (eds) A Companion to Cultural Geography. Blackwell Press, Oxford.

Johnston, C. and Pratt, G. (2010) Nanay (mother): A testimonial play. cultural geographies, 17, pp. 123–133.

Jones, O. (2009) Dwelling. In Kitchin, R. and Thrift, N. (eds) International Encyclopedia of Human Geography. Elsevier, Amsterdam.

Jones, P. (2005) Performing the city: A body and a bicycle take on Birmingham, UK. Social & Cultural Geography, 6, pp. 813–830.

Kanngieser, A. (2012) A sonic geography of voice: Towards an affective politics. Progress in Human Geography, 36, pp. 336–353.

Keating, T.P. (2019a) Pre-individual affects: Gilbert Simondon and the individuation of relation. cultural geographies, 26, pp. 211–226.

Keating, T.P. (2019b) Imaging. *Transactions of the Institute of British Geographers*, 44, pp. 654–656.

Kenzer, M.S. (1985) Milieu and the 'intellectual landscape': Carl O. Sauer's undergraduate heritage. *Annals of the Association of American Geographers*, 75, pp. 258–270.

Kershaw, B. (2007) *Theatre Ecology: Environments and Performance Events*. Cambridge University Press, Cambridge.

Kinsley, S. (2012) Futures in the making: Practices to anticipate 'ubiquitous computing'. *Environment and Planning A*, 44, pp. 1554–1569.

Kinsley, S. (2014) The matter of 'virtual' geographies. *Progress in Human Geography*, 38, pp. 364–384.

Kinsley, S. (2015) Memory programmes: The industrial retention of collective life. *cultural geographies*, 22, pp. 155–175.

Kirch, S. (2013) Cultural geography 1: Materialist turns. *Progress in Human Geography*, 37, pp. 433–441.

Kneale, J. (2006) From beyond: H. P. Lovecraft and the place of horror. *cultural geographies*, 13, pp. 106–126.

Kniffen, F. (1965) Folk housing: Key to diffusion. *Annals of the Association of American Geographers*, 55, pp. 549–577.

Kraftl, P. and Adey, P. (2008) Architecture/affect/inhabitation: Geographies of being-in buildings. *Annals of the Association of American Geographers*, 98, pp. 213–231.

Lapworth, A. (2015) Habit, art, and the plasticity of the subject: The ontogenetic shock of the bioart encounter. *cultural geographies*, 22, pp. 85–102.

Latham, A. (2003a) Research, performance, and doing human geography: Some reflections on the diary-photograph, diary-interview method. *Environment and Planning A*, 35, pp. 1993–2017.

Latham, A. (2003b) The possibilities of performance. *Environment and Planning A*, 35, pp. 1901–1906.

Latham, A. (2004) Researching and writing everyday accounts of the city: An introduction to the diary-photo diary interview method. In Knowles, C. and Sweetman, P. (eds) *Picturing the Social Landscape: Visual Methods and the Sociological Imagination*. Routledge, London.

Latham, A. and McCormack, D. (2004) Moving cities: Rethinking the materialities of urban geographies. *Progress in Human Geography*, 28, pp. 701–724.

Latham, A. and McCormack, D. (2009) Thinking with images in non-representational cities: Vignettes from Berlin. *Area*, 41, pp. 252–262.

Latham, A. and Wood, P.R.H. (2015) Inhabiting infrastructure: Exploring the interactional spaces of urban cycling. *Environment and Planning A*, 47, pp. 300–319.

Latour, B. (2005) *Reassembling the Social: An Introduction to Actor-network-theory*. Oxford University Press, Oxford.

Laurier, E. (2001) Why people say where they are during mobile phone calls. *Environment and Planning D: Society and Space*, 19, pp. 485–504.

Laurier, E. (2005) Searching for a parking space. *Intellectica*, 2–3, pp. 101–116.

Laurier, E. (2010) Representation and everyday use: How to feel things with words. In Anderson, B. and Harrison, P. (eds) *Taking-place: Non-representational Theories and Geography*. Ashgate, Aldershot.

Laurier, E. (2016) YouTube: Fragments of a video-tropic atlas. *Area*, 48, pp. 488–495.

Laurier, E., Maze, R. and Lundin, J. (2006) Putting the dog back in the park: Animal and human mind-in-action. *Mind, Culture and Activity*, 13, pp. 2–24.

Laurier, E. and Philo, C. (1999) X-morphising: Review essay of Bruno Latour's *Aramis, or the Love of Technology*. *Environment and Planning A*, 31, pp. 1047–1071.

Laurier, E. and Philo, C. (2006a) Cold shoulders and napkins handed: Gestures of responsibility. *Transactions of the Institute of British Geographers*, 31, pp. 193–208.

Laurier, E. and Philo, C. (2006b) Possible geographies: A passing encounter in a café. *Area*, 38, pp. 353–363.

Laurier, E. and Philo, C. (2006c) Natural problems of naturalistic video data. In Knoblauch, H., Raab, J., Soeffner, H.-G. and Schnettler, B. (eds.) *Video-analysis Methodology and Methods, Qualitative Audiovisual Data Analysis in Sociology*. Peter Lang, Oxford.

Lea, J. (2009) Post-phenomenological geographies. In Kitchen, R. and Thrift, N. (eds) *International Encyclopedia of Human Geography*. Elsevier, London.

Lea, J., Cadman, L. and Philo, C. (2015) Changing the habits of a lifetime? Mindfulness meditation and habitual geographies. *cultural geographies*, 22, pp. 49–65.

Lees, L. (2002) Rematerializing geography: The 'new' urban geography. *Progress in Human Geography*, 26, pp. 101–112.

Lefebvre, H. (1991) *The Production of Space*. Blackwell, London.

Lefebvre, H. (1992) *Critique of Everyday Life Volume I: Introduction*. Verso, London.

Lefebvre, H. (2002) *Critique of Everyday Life Volume II: Foundations for a Sociology of the Everyday*. Verso, London.

Lefebvre, H. (2004) *Rhythmanalysis: Space, Time and Everyday Life*. Continuum, London.

Lefebvre, H. (2005) *Critique of Everyday Life Volume III: From Modernity to Modernism (Towards a Metaphilosophy of Daily Life)*. Verso, London.

Lefebvre, H. and Regulier, C. (2004) The rhythmanalytical project. In Lefebvre, H. *Rhythmanalysis: Space, Time and Everyday Life*. Continuum, London.

Leonard, M. (2005) Performing identities: Music and dance in the Irish communities of Coventry and Liverpool. *Social & Cultural Geography*, 6, pp. 515–529.

Ley, D. (2009) Lifeworld. In Gregory, D., Johnston, R., Pratt, G., Watts, M.J. and Whatmore, S. (eds) *The Dictionary of Human Geography*. Wiley-Blackwell, Oxford.

Ley, D. and Samuels, M.S. (eds) (1978) *Humanistic Geography: Prospects and Problems*. Croom Helm, London.

Leyshon, A., Matless, D. and Revill, G. (eds) (1998) *The Place of Music*. The Guildford Press, London.

Lim, J. (2010) Immanent politics: Thinking race and ethnicity through affect and machinism. *Environment and Planning A*, 42, pp. 2393–2409.

Lingis, A. (2004) The music of space. In Foltz, B.V. and Frodeman, R. (eds) *Rethinking Nature: Essay in Environmental Philosophy*. Indiana University Press, Bloomington.

Longhurst, R. (2001) *Bodies: Exploring Fluid Boundaries*. Routledge, London.

Lorimer, H. (2005) Cultural geography: The busyness of being 'more-than-representational'. *Progress in Human Geography*, 29, pp. 83–94.

Lorimer, H. (2006) Herding memories of humans and animal. *Environment and Planning D: Society and Space*, 24, pp. 497–518.

Lorimer, H. (2007) Cultural geography: Worldly shapes, differently arranged. *Progress in Human Geography*, 31, pp. 89–100.

Lorimer, H. (2008) Cultural geography: Non-representational conditions and concerns. *Progress in Human Geography*, 32, pp. 551–559.

Lorimer, H. (2015) Afterword: Non-representational theory and me too. In Vannini, P. (ed) *Non-representational Methodologies: Re-envisioning Research*. Routledge, London.

Lorimer, H. and Lund, K. (2008) A collectable topography: Walking, remembering and recording mountains. In Ingold, T. and Vergunst, J.L. (eds.) *Ways of Walking: Ethnography and Practice on Foot*. Ashgate, Aldershot.

Lorimer, H. and Wylie, J. (2010) LOOP (a geography). *Performance Research: A Journal of the Performing Arts*, 15, pp. 6–13.

Lorimer, J. (2010) Moving image methodologies for more-than-human geographies *cultural geographies*, 17, pp. 237–258

Maclaren, A.S. (2018) Affective lives of rural aging. *Sociologia Ruralis*, 58, pp. 213–234.

Macpherson, H. (2008) "I don't know why they call it the Lake District they might as well call it the rock district!" The workings of humour and laughter in research with members of visually impaired walking groups. *Environment and Planning D: Society and Space*, 26, pp. 1080–1095.

Macpherson, H. (2009) The inter-corporeal emergence of landscape: Negotiating sight, blindness and ideas of landscape in the British countryside. *Environment and Planning A*, 41, pp. 1042–1054.

Macpherson, H. (2010) Non-representational approaches to body-landscape relations. *Geography Compass*, 4, pp. 1–13.

Maddern, J. and Adey, P. (2008) Editorial: Spectro-geographies. *cultural geographies*, 15, pp. 291–295.

Maddrell, A. and della Dora, V. (2013) Crossing surfaces in search of the holy: Landscape and liminality in contemporary Christian pilgrimage. *Environment and Planning A*, 45, pp. 1105–1126.

Malbon, B. (1999) *Clubbing: Dancing, ecstasy, vitality*. Routledge, London.

Martin, C. (2011) Fog-bound: Aerial space and the elemental entanglements of body-with-world. *Environment and Planning D: Society and Space*, 29, pp. 454–468.

Massey, D. (2005) *For Space*. Sage, London.

Massumi, B. (2002) *Parables of the Virtual: Movement, Affect, Sensation*. Duke University Press, London.

Mathewson, K. (2011) Sauer's Berkeley School legacy: Foundation for an emergent environmental geography? in Bocco, G., Urquijo, P.S. and Vieyra, A. (eds) *Geografía y Ambiente en América Latina*. CIGA/UNAM, Morelia.

Matless, D. (1998) *Landscape and Englishness*. Reaktion, London.

Matless, D. (2005) Sonic geography in a nature region. *Social & Cultural Geography*, 6, pp. 745–766.

McCormack, D.P. (2002) A paper with an interest in rhythm. *Geoforum*, 33, pp. 469–485.

McCormack, D.P. (2003) An event of geographical ethics in spaces of affect. *Transactions of the Institute of British Geographers*, 28, pp. 488–507.

McCormack, D.P. (2005) Diagramming practice and performance. *Environment and Planning D: Society and Space*, 23, pp. 119–147.

McCormack, D.P. (2006) For the love of pipes and cables: A response to Deborah Thien. *Area*, 38, pp. 330–332.

McCormack, D.P. (2007) Molecular affects in human geographies. *Environment and Planning A*, 39, pp. 359–377.

McCormack, D.P. (2008a) Engineering affective atmospheres on the moving geographies of the 1897 Andree expedition. *cultural geographies*, 15, pp. 413–430.

McCormack, D.P. (2008b) Geographies for moving bodies: Thinking, dancing, spaces. *Geography Compass*, 2, pp. 1822–1836.

McCormack, D.P. (2012) Geography and abstraction: Towards an affirmative critique. *Progress in Human Geography*, 36, pp. 715–734.

McCormack, D.P. (2013) *Refrains for Moving Bodies: Experience and Experiment in Affective Spaces.* Duke University Press, Durham.

McCormack, D.P. (2014) Atmospheric things and circumstantial excursions. *cultural geographies*, 21, pp. 605–625.

McCormack, D.P. (2015a) Governing inflation: price and atmospheres of emergency. *Theory, Culture and Society*, 32, pp. 131–154.

McCormack, D.P. (2015b) Envelopment, exposure, and the allure of becoming elemental. *Dialogues in Human Geography*, 5, pp. 85–89.

McCormack, D.P. (2015c) Devices for doing atmospheric things. In Vannini, P. (ed) *Non-representational Methodologies: Re-envisioning Research*. Routledge, London.

McCormack, D.P. (2017) The circumstances of post-phenomenological life worlds. *Transactions of the Institute of British Geographers*, 42, pp. 2–13.

McKenzie, J. (1998) Gender trouble: (The) Butler did it. In Phelan, P. and Lave, K. (eds) *The Ends of Performance*. New York University Press, London.

Meehan, K., Shaw, I.G.R. and Marston, S.A. (2013) Political geographies of the object. *Political Geography*, 33, pp. 1–10.

Meehan, K., Shaw, I.G.R. and Marston, S.A. (2014) The state of objects. *Political Geography*, 39, pp. 60–62.

Mels, T. (ed) (2004) *Reanimating Places: A Geography of Rhythms*. Ashgate, Aldershot.

Merriman, P. (2014) Rethinking mobile methods, *Mobilities*, 9, pp. 167–187.

Merriman, P., Revill, G., Cresswell, T., Lorimer, H., Matless, D., Rose, G. and Wylie, J. (2008) Landscape, mobility, practice. *Social & Cultural Geography*, 9, pp. 191–212.

Mitchell, D. (1996) *The Lie of the Land: Migrant Workers and the California Landscape*. University of Minnesota Press, Minneapolis.

Mitchell, D. (2000) *Cultural Geography: A Critical Introduction*. Blackwell, Oxford.

Mitchell, D. (2003) *The Right to the City: Social Justice and the Fight for Public Space*. The Guildford Press, London.

Mitchell, K. and Elwood, S. (2012) Mapping children's politics: The promise of articulation and the limits of nonrepresentational theory. *Environment and Planning D: Society and Space*, 30, pp. 788–804.

Morton, F. (2005) Performing ethnography: Irish traditional music sessions and new methodological spaces. *Social & Cultural Geography*, 6, pp. 661–676.

Mott, C. and Cockayne, D. (2017) Citation matters: Mobilizing the politics of citation toward a practice of 'conscientious engagement'. *Gender, Place and Cultural: A Journal of Feminist Geography*, 24, pp. 954–973.

Müller, M. and Schurr, C. (2016) Assemblage thinking and actor-network theory: Conjunctions, disjunctions, cross-fertilisations. *Transactions of the Institute of British Geographers*, 41, pp. 217–229.

Murdock, J. (1997) Inhuman/nonhuman/human: Actor-network theory and the prospects for a nondualistic and symmetrical perspective on nature and society. *Environment and Planning D: Society and Space*, 15, pp. 731–756.

Murdock, J. (1998) The spaces of actor-network theory. *Geoforum*, 29, pp. 357–374.

Nancy, J.-L. (2007) *Listening*. Fordham University Press, New York.

Nash, C. (2000) Performativity in practice: Some recent work in cultural geography. *Progress in Human Geography*, 24, pp. 653–664.

Nayak, A. and Jeffrey, A. (2011) *Geographical Thought: An Introduction to Ideas in Human Geography*. Prentice Hall, Harlow.

Noxolo, P. (2018) Flat out! Dancing the city at a time of austerity. *Environment and Planning D: Society and Space*, 36, pp. 797–811.

O'Grady, N. (2018) Geographies of affect. *Oxford Bibliographies*. Available at: http://www.oxfordbibliographies.com/view/document/obo-9780199874002/obo-9780199874002-0186.xml

Parkes, D.N. and Thrift, N. (1980) *Times, Spaces, and Places: A Chronogeographic Perspective*. John Wiley and Sons, Chichester.

Patchett (2010) *A Rough Guide to Non-representational Theory*. Available at: https://merlepatchett.wordpress.com/2010/11/12/a-rough-guide-to-non-representational-theory/

Paterson, M. (2009) Haptic geographies: Ethnography, haptic knowledges and sensuous dispositions. *Progress in Human Geography*, 33, pp. 766–788.

Paterson, M. and Glass, M.R. (2020) Seeing, feeling, and showing 'bodies-in-place': Exploring reflexivity and the multisensory body through videography. *Social & Cultural Geography*, 21, pp. 1–24.

Pavia, D. (2018) Dissonance: Scientific paradigms undermining the study of sound in geography. *Fennia*, 196, pp. 77–87.

Pearson, M. (2006) *«In comes I»: Performance, Memory and Landscape*. University of Exeter Press, Exeter.

Phelan, P. (1993) *Unmarked: The Politics of Performance*. Routledge, London.

Phelan, P. (1997) *Mourning Sex: Performing Public Memories*. Routledge, London.

Philo, C. (2000) More words, more worlds: Reflections on the cultural turn and human geography. In Cook, I., Crouch, D., Naylor, S. and Ryan, J. (eds) *Cultural Turns/Geographical Turns: Perspectives on Cultural Geography*. Pearson Education, Harlow.

Pinder, D. (2001) Ghostly footsteps: Voices, memories and walks in the city. *Ecumene*, 8, pp. 1–19.

Price, M. and Lewis, M. (1993) The reinvention of cultural geography. *Annals of the Association of American Geographers*, 83, pp. 1–17.

Raynor, R. (2017a) Dramatising austerity: Holding a story together (and why it falls apart...). *cultural geographies*, 24, pp. 193–212.

Raynor, R. (2017b) (De)composing habit in theatre-as-method. *GeoHumanities*, 3, pp. 108–121.

Raynor, R. (2019) Speaking, feeling, mattering: Theatre as method and model for practice-based, collaborative, research. *Progress in Human Geography*, 43, pp. 691–710.

Revill, G. (2000a) English pastoral: Music, landscape, history and politics. In Cook, I., Crouch, D., Naylor, S. and Ryan, J. (eds) *Cultural Turns/Geographical Turns: Perspectives on Cultural Geography*. Routledge, London.

Revill, G. (2000b) Music and the politics of sound: Nationalism, citizenship, and auditory space. *Environment and Planning D: Society and Space*, 18, pp. 597–613.

Revill, G. (2004) Performing French folk music: Dance, authenticity and nonrepresentational theory. *cultural geographies*, 11, pp. 199–209.

Revill, G. (2016) How is space made in sound? Spatial mediation, critical phenomenology and the political agency of sound. *Progress in Human Geography*, 40, pp. 240–256.

Robbins, P. and Marks, B. (2009) Assemblage geographies. In Smith, S.J., Pain, R., Marston, S.A. and Jones III, J.P. (eds) *The SAGE Handbook of Social Geographies*. Sage, London.

Roberts, T. (2012) From 'new materialism' to 'machinic assemblage': Agency and affect in IKEA. *Environment and Planning A*, 44, pp. 2512–2529.

Roberts, T. (2019a) Resituating post-phenomenological geographies: Deleuze, relations and the limits of objects. *Transactions of the Institute of British Geographers*, 44, pp. 542–554.

Roberts, T. (2019b) Writing. *Transactions of the Institute of British Geographers*, 44, pp. 644–646.

Roe, E.J. (2006) Material connectivity, the immaterial and the aesthetic of eating practices: An argument for how genetically modified foodstuff becomes inedible. *Environment and Planning A*, 38, pp. 465–481.

Roe, E.J. (2010) Ethics and the non-human: The matterings of animal sentience in the meat industry. In Anderson, B. and Harrison, P. (eds) *Taking-place: Non-representational Theories and Geography*. Ashgate, Farnham.

Rogers, A. (2010) Geographies of performing scripted language. *cultural geographies*, 17, pp. 53–75.

Rogers, A. (2012a) Geographies of the performing arts: Landscapes, places and cities. *Geography Compass*, 6, pp. 60–75.

Rogers, A. (2012b) Emotional geographies of method acting in Asian American theater. *Annals of the Association of American Geographers*, 102, pp. 423–442.

Romanillos, J.L. (2008) 'Outside, it is snowing': Experience and finitude in the non-representational landscapes of Alain Robbe-Grillet. *Environment and Planning D: Society and Space*, 26, pp. 795–822.

Romanillos, J.L. (2011) Geography, death and finitude. *Environment and Planning A*, 43, pp. 2533–2553.

Romanillos, J.L. (2015) Mortal questions: Geographies on the other side of life. *Progress in Human Geography*, 39, pp. 560–579.

Rose, G. (1993) *Feminism and Geography*. Polity Press, Cambridge.

Rose, M. (2006) 'Gathering dreams of presence': A project for the cultural landscape. *Environment and Planning D: Society and Space*, 24, pp. 537–554.

Rose, M. (2010a) Envisioning the future: Ontology, time and the politics of non-representation. In Anderson, B. and Harrison, P. (eds) *Taking-place: Non-representational Theories and Geography*. Ashgate, London.

Rose, M. (2010b) Pilgrims: An ethnography of sacredness. *cultural geographies*, 17, pp. 507–524.

Rose, M. (2012) Dwelling as marking and claiming. *Environment and Planning D: Society and Space*, 30, pp. 757–771.

Rose, M. (2016) A place for other stories: Authorship and evidence in experimental times. *Geohumanities*, 2, pp. 132–148.

Rose, M. and Wylie, J. (2006) Animating landscape. *Environment and Planning D: Society and Space*, 24, pp. 475–479.

Rosenstein, B. (2002) Video use in social science research and program evaluation *International Journal of Qualitative Methods*, 1, pp. 22–43.

Saldanha, A. (2005) Trance and visibility at dawn: Racial dynamics in Goa's rave scene. *Social & Cultural Geography*, 6, pp. 707–721.

Saldanha, A. (2007) *Psychedelic White: Goa Trance and the Viscosity of Race.* University of Minnesota Press, Minneapolis.

Saldahna, A. (2009) Soundscapes. In Kitchen, R. and Thrift, N. (eds) *International Encyclopedia of Human Geography*, Elsever.

Saldanha, A. (2010a) Skin, affect, aggregation: Guattarian variations of Fannon. *Environment and Planning A*, 42, pp. 2410–2427.

Saldanha, A. (2010b) Politics and difference. In Anderson, B. and Harrison, P. (eds) *Taking-place: Non-representational Theories and Geography*. Ashgate, London.

Sauer, C. (2008) The morphology of the landscape. In Oakes, T.S. and Price, P.L. (eds) *The Cultural Geography Reader*. Routledge, London.

Schatzki, T.R. (2001) Introduction: Practice theory. In Schatzki, T.R., Cetina, K.K. and von Savigny, E. (eds) *The Practice Turn in Contemporary Theory*. Routledge, London.

Schechner, R. (2002) *Performance Studies: An Introduction.* Routledge, London.

Schechner, R. (2003) *Performance Theory.* Routledge, London.

Schusterman, R. (2000) *Performing Live: Aesthetic Alternatives for the Ends of Art.* Cornell University Press, London.

Scott, H. (2004) Cultural turns. In Duncan, J.S., Johnson, N.C. and Schein, R.H. (eds) *A Companion to Cultural Geography*. Blackwell, Oxford.

Scriven, R. (2014) Geographies of pilgrimage: Meaningful movements and embodied mobilities. *Geography Compass*, 8, pp. 249–261.

Seamon, D. (1980) Body-subject, time-space routines, and place-ballets. In Buttimer, A. and Seamon, D. (eds) *The Human Experience of Space and Place*. Croom Helm, London.

Seamon, D. (1993) Dwelling, seeing and designing: An introduction. In Seamon, D. (ed) *Dwelling, Seeing, and Designing: Towards a Phenomenological Ecology.* SUNY Press, Albany.

Seigworth, G.J. and Gregg, M. (eds) (2010a) *The Affect Theory Reader.* Duke University Press, London.

Seigworth, G.J. and Gregg, M. (2010b) An inventory of shimmers. In Seigworth, G.J. and Gregg, M. (eds) *The Affect Theory Reader*. Duke University Press, London.

Semple, E.C. (1911) *Influences of Geographic Environment.* Available at: https://archive.org/details/influencesofgeog00semp

Shaw, I.G.R. (2012) Towards an evental geography. *Progress in Human Geography*, 36, pp. 613–627.

Shaw, I.G.R. and Meehan, K. (2013) Force-full: Power, politics and object-oriented philosophy. *Area*, 45, pp. 216–222.

Shaw, J. and Docherty, I. (2014) *The Transport Debate.* Policy Press, Bristol.

Shaw, R. (2014) Beyond night-time economy: Affective atmospheres of the urban night. *Geoforum*, 51, pp. 87–95.

Sidaway, J.D. (2009) Shadows on the path: Negotiating geopolitics on an urban section of Britain's South West Coast Path. *Environment and Planning D: Society and Space*, 27, pp. 1091–1116.

Simonsen, K. (2004) Spatiality, temporality and the construction of the city. In Baerenholdt, J.O. and Simonsen, K. (eds) *Space Odysseys: Spatiality and Social Relations in the 21st Century*. Ashgate, Aldershot.

Simonson, K. (2005) Bodies, sensations, space and time: The contribution from Henri Lefebvre. *Geografiska Annaler. Series B. Human Geography*, 87, pp. 1–14.

Simonsen, K. (2013) In quest of a new humanism: Embodiment, experience and phenomenology as critical geography. *Progress in Human Geography*, 37, pp. 10–26.

Simpson, P. (2008) Chronic everyday life: Rhythmanalysing street performance. *Social & Cultural Geography*, 9, pp. 807–829.

Simpson, P. (2009) 'Falling on deaf ears': A post-phenomenology of sonorous presence. *Environment and Planning A*, 41, pp. 2556–2575.

Simpson, P. (2011a) Street performance and the city: Public space, sociality, and intervening in the everyday. *Space and Culture*, 14, pp. 415–430.

Simpson, P. (2011b) 'So, as you can see . . .': Some reflections on the utility of video methodologies in the study of embodied practices. *Area*, 43, pp. 343–352.

Simpson, P. (2012) Apprehending everyday rhythms: Rhythmanalysis, time-lapse photography, and the space-times of street performance. *cultural geographies*, 19, pp. 423–445.

Simpson, P. (2013) Ecologies of experience: Materiality, sociality, and the embodied experience of (street) performing. *Environment and Planning A*, 45, pp. 180–196.

Simpson, P. (2014a) A soundtrack to the everyday: Street music and the production of convivial 'healthy' public places. In Andrews, G., Kingsbury, P. and Kearns, R. (eds) *Soundscapes of Wellbeing in Popular Music*. Ashgate, Aldershot.

Simpson, P. (2014b) Spaces of affect. In Adams, P., Craine, J. and Dittmer, J. (eds) *Ashgate Research Companion on Geographies of Media*. Ashgate, Aldershot.

Simpson, P. (2015a) Nonrepresentational theory. In Warf, B. (ed) *Oxford Bibliographies in Geography*. Oxford University Press, New York.

Simpson, P. (2015b) What remains of the intersubjective? On the presencing of self and other. *Emotion, Space and Society*, 14, pp. 65–73.

Simpson, P. (2017a) Non-representational theory. In *The Wiley-AAG International Encyclopedia of Geography: People, the Earth, Environment, and Technology*.

Simpson, P. (2017b) A sense of the cycling environment: Felt experiences of infrastructure and atmospheres. *Environment and Planning A*, 49, pp. 426–447.

Simpson, P. (2017c) Sonic affects and the production of space: 'Music by handle' and the politics of street music in Victorian London. *cultural geographies*, 24, pp. 89–109.

Simpson, P. (2017d) Spacing the subject: Thinking subjectivity after non-representational theory. *Geography Compass*, 11, p. e12347.

Simpson, P. (2019) Elemental mobilities: Atmospheres, matter and cycling amid the weather-world. *Social & Cultural Geography*, 20, pp. 1050–1069.

Smith, D.P. and Hubbard, P. (2014) The segregation of educated youth and dynamic geographies of studentification. *Area*, 46, pp. 92–100.

Smith, R.G. (2003) Baudrillard's non-representational theory: Burn the signs and journey without maps. *Environment and Planning D: Society and Space,* 21, pp. 67–84.

Smith, S.J. (1984) Practicing humanistic geography. *Annals of the Association of American Geographers,* 74, pp. 353–374.

Smith, S.J. (1994) Soundscape. *Area,* 26, pp. 232–240.

Smith, S.J. (2000) Performing the (sound) world. *Environment and Planning D: Society and Space,* 18, pp. 615–637.

Solnit, R. (2001) *Wanderlust: A History of Walking.* Verso, London.

Solot, M. (1986) Carl Sauer and cultural evolution. *Annals of the Association of American Geographers,* 76, pp. 508–520.

Spinney, J. (2009) Cycling the city: Movement, meaning and method. *Geography Compass,* 3, pp. 817–835.

Stagoll, C. (2005) Event. In Parr, A. (ed) *The Deleuze Dictionary.* Edinburgh University Press, Edinburgh.

Stewart, K. (2007) *Ordinary Affects.* Duke University Press, London.

Stewart, K. (2010) Worlding refrains. In Seigworth, G.J. and Gregg, M. (eds) *The Affect Theory Reader.* Duke University Press, London.

Sumartojo, S. and Pink, S. (2019) *Atmospheres and the Experiential World: Theory and Methods.* Routledge, Abingdon.

Swanton, D. (2010) Sorting bodies: Race, affect and everyday multiculture in a mill-town in Northern England. *Environment and Planning A,* 42, pp. 2232–2250.

Tanenbaum, S.J. (1995) *Underground Harmonies: Music and Politics in the Subways of New York.* Cornell University Press, London.

Thibaud, J.-P. (2015) The backstage of urban ambiances: When atmospheres pervade everyday experience. *Emotion, Space and Society,* 15, pp. 39–46.

Thien, D. (2005) After or beyond feeling? A consideration of affect and emotion in geography. *Area,* 37, pp. 450–454.

Thornton, P. (2015) The meaning of light: Seeing and being on the battlefield. *Cultural Geographies,* 22, pp. 567–583.

Thrift, N. (1977) *An Introduction to Time-Geography.* Geoabstracts Ltd., Norwich.

Thrift, N. (1996) *Spatial Formations.* Sage, London.

Thrift, N. (1997) The still point. In Pile, S. and Keith, M. (Eds.) *Geographies of Resistance.* Routledge, London.

Thrift, N. (1999) Steps to an ecology of place. In Massey, D., Allen, J. and Sarre, P. (eds) *Human Geography Today.* Polity, Cambridge.

Thrift, N. (2000) Afterwords. *Environment and Planning D: Society and Space,* 18, pp. 213–255.

Thrift, N. (2003) Performance and …. *Environment and Planning A,* 35, pp. 2019–2024.

Thrift, N. (2004a) Driving in the city. *Theory, Culture and Society,* 21, pp. 41–59.

Thrift, N. (2004b) Intensities of feeling: Towards a spatial politics of affect. *Geografiska Annaler B,* 86, pp. 57–78.

Thrift, N. (2004c) Remembering the technological unconscious by foregrounding knowledges of position. *Environment and Planning D: Society and Space,* 22, pp. 175–190.

Thrift, N. (2008) *Non-representational Theory: Space, Politics, Affect.* Routledge, London.

Thrift, N. (2010) Understanding the material practices of glamour. In Seigworth, G.J. and Gregg, M. (eds) *The Affect Theory Reader.* Duke University Press, London.

Thrift, N. and Dewsbury, J.-D. (2000) Dead geography – And how to make them live. *Environment and Planning D: Society and Space*, 18, pp. 411–432.

Thrift, N., Harrison, P. and Anderson, B. (2010) 'The 27th Letter': An interview with Nigel Thrift. In Anderson, B. and Harrison, P. (eds) *Taking-place: Non-representational Theories and Geography*. Ashgate, Aldershot.

Till, K. (2005) *The New Berlin: Memory, Politics, Place*. University of Minnesota Press, Minneapolis.

Tolia-Kelly, D.P. (2004) Materializing post-colonial geographies: Examining the textural landscapes of migration in the South Asian home. *Geoforum*, 35, pp. 675–688.

Tolia-Kelly, D.P. (2006a) Affect – An ethnocentric encounter? Exploring the 'universalist' imperative of emotional/affectual geographies. *Area*, 38, pp. 213–217.

Tolia-Kelly, D.P. (2006b) Mobility/stability: British Asian cultures of 'Landscape and Englishness'. *Environment and Planning A*, 38, pp. 341–358.

Tolia-Kelly, D.P. (2011) The geographies of cultural geography III: Material geographies, vibrant matters and risking surface geographies. *Progress in Human Geography*, 37, pp. 153–160.

Vannini, P. (ed) (2015a) *Non-representational Methodologies: Re-envisioning Research*. Routledge, London.

Vannini, P. (2015b) Non-representational research methodologies: An introduction. In Vannini, P. (ed) *Non-representational Methodologies: Re-envisioning Research*. Routledge, London.

Vannini, P. (2015c) Non-representational ethnography: New ways of animating lifeworlds. *cultural geographies*, 22, pp. 317–327.

Vannini, P. and Stewart, L.M. (2017) The GoPro gaze. *cultural geographies*, 24, pp. 149–155.

Vannini, P., Waskul, D., Gottschalk, S. and Ellis-Newstead, T. (2012) Making sense of the weather: Dwelling and weathering on Canada's rain coast. *Space and Culture*, 15, pp. 261–380.

Veale, L., Endfield, G. & Naylor, S. (2014) Knowing weather in place: The Helm Wind of Cross Fell. *Journal of Historical Geography*, 45, pp. 25–37.

Waitt, G., Harada, T. and Duffy, M. (2017) 'Let's have some music': Sound, gender and car mobility. *Mobilities*, 12, pp. 324–342.

Wallach, B. (1999) Commentary: Will Carl Sauer make it across that great bridge to the next millennium? *Yearbook of the Association of Pacific Coast Geographers*, 61, pp. 129–136.

Whatmore, S. (2002) *Hybrid Geographies: Natures, Cultures, Space*. Sage, London.

Whatmore, S. (2006) Materialist returns: Practising cultural geography in and for a more-than-human world. *cultural geographies*, 13, pp. 600–609.

Whatmore, S. and Thorne, L. (2000) Elephants on the move: Spatial formations of wildlife exchange. *Environment and Planning D: Society and Space*, 18, pp. 185–203.

Whyte, W.H. (1980) *The Social Life of Small Urban Space*. The Project for Public Space, New York.

Williams, N. (2016) Creative processes: From interventions in art to intervallic experiments through Bergson. *Environment and Planning A*, 48, pp. 1549–1564.

Williams, N. (2017) *An Aesthetic Gait: Research in the Minor Registers of Creativity and Walking*. Unpublished PhD Thesis, School of Geographical Sciences, University of Bristol.

Wilson, H.F. (2011) Passing propinquities in the multicultural city: The everyday encounters of bus passengering. *Environment and Planning A*, 43, pp. 634–649.

Wilson, H.F. (2013) Learning to think differently: Diversity training and the 'good encounter'. *Geoforum*, 45, pp. 73–82.

Wilson, H.F. (2014) The possibilities of tolerance: Intercultural dialogue in a multi-cultural Europe. *Environment and Planning D: Society and Space*, 32, pp. 852–868.

Wilson, H.F. (2017) On geography and encounter. *Progress in Human Geography*, 41, pp. 451–471.

Wilson, M. (2011a) 'Training the eye': Formation of the geocoding subject. *Social & Cultural Geography*, 12, pp. 357–376.

Wilson, M. (2011b) Data matter(s): Legitimacy, coding, and qualifications-of-life. *Environment and Planning D: Society and Space*, 29, pp. 857–872.

Wilson, M. (2015) Paying attention, digital media, and community-based critical GIS. *cultural geographies*, 22, pp. 177–191.

Wood, N. (2012) Playing with 'Scottishness': Musical performance, non-representational thinking and the 'doings' of national identity. *cultural geographies*, 19, pp. 195–215.

Wood, N. and Smith, S.J. (2004) Instrumental routes to emotional geographies, *Social & Cultural Geography*, 5, pp. 533–548.

Wood, N., Duffy, M. and Smith, S.J. (2007) The art of doing (geographies of) music. *Environment and Planning D: Society and Space*, 25, pp. 867–889.

Wylie, J. (2002a) An essay on ascending Glastonbury Tor. *Geoforum*, 33, pp. 441–454.

Wylie, J. (2002b) Becoming-icy: Scott and Amundsens's polar voyages. *cultural geographies*, 9, pp. 249–265.

Wylie, J. (2003) Landscape, performance and dwelling: A Glastonbury case study. In Cloke, P. (ed) *Country Visions*. Pearson, Harlow.

Wylie, J. (2005) A single day's walking: Narrating self and landscape on the South West Coast Path. *Transactions of the Institute of British Geographers*, 30, pp. 234–247.

Wylie, J. (2006a) Poststructuralist theories, critical methods and experimentation. In Aitken, S. and Valentine, G. (eds) *Approaches to Human Geography*. Sage, London.

Wylie, J. (2006b) Depths and folds: On landscape and the gazing subject. *Environment and Planning D Society and Space*, 24, pp. 519–535.

Wylie, J. (2006c) Smoothlands: Fragments/landscapes/fragments. *cultural geographies*, 13, pp. 458–465.

Wylie, J. (2007a) *Landscape*. Routledge, London.

Wylie, J. (2007b) The spectral geographies of W.G. Sebald. *cultural geographies*, 14, pp. 171–188.

Wylie, J. (2009) Landscape, absence and the geographies of love. *Transactions of the Institute of British Geographers*, 34, pp. 275–289.

Wylie, J. (2010) Non-representational Subjects? In Anderson, B. and Harrison, P. (eds) *Taking-place: Non-representational Theories and Geography*. Ashgate, Aldershot.

Wylie, J. (2012) Dwelling and displacement: Tim Robinson and the questions of landscape. *cultural geographies*, 19, pp. 365–383.

Yorgason, E. and della Dora, V. (2009) Editorial: Geography, religion, and emerging paradigms: Problematizing the dialogue. *Social & Cultural Geography*, 10, pp. 629–637.

INDEX

Note: Page numbers followed by "n" denote endnotes.

Printed in the United States
by Baker & Taylor Publisher Services